AF864051

Springer Theses

Recognizing Outstanding Ph.D. Research

Aims and Scope

The series "Springer Theses" brings together a selection of the very best Ph.D. theses from around the world and across the physical sciences. Nominated and endorsed by two recognized specialists, each published volume has been selected for its scientific excellence and the high impact of its contents for the pertinent field of research. For greater accessibility to non-specialists, the published versions include an extended introduction, as well as a foreword by the student's supervisor explaining the special relevance of the work for the field. As a whole, the series will provide a valuable resource both for newcomers to the research fields described, and for other scientists seeking detailed background information on special questions. Finally, it provides an accredited documentation of the valuable contributions made by today's younger generation of scientists.

Theses may be nominated for publication in this series by heads of department at internationally leading universities or institutes and should fulfill all of the following criteria

- They must be written in good English.
- The topic should fall within the confines of Chemistry, Physics, Earth Sciences, Engineering and related interdisciplinary fields such as Materials, Nanoscience, Chemical Engineering, Complex Systems and Biophysics.
- The work reported in the thesis must represent a significant scientific advance.
- If the thesis includes previously published material, permission to reproduce this must be gained from the respective copyright holder (a maximum 30% of the thesis should be a verbatim reproduction from the author's previous publications).
- They must have been examined and passed during the 12 months prior to nomination.
- Each thesis should include a foreword by the supervisor outlining the significance of its content.
- The theses should have a clearly defined structure including an introduction accessible to new PhD students and scientists not expert in the relevant field.

Indexed by zbMATH.

More information about this series at http://www.springer.com/series/8790

Antonios M. Alvertis

On Exciton–Vibration and Exciton–Photon Interactions in Organic Semiconductors

Doctoral Thesis accepted by
University of Cambridge, United Kingdom

Springer

Author
Dr. Antonios M. Alvertis
Cavendish Laboratory
University of Cambridge
Cambridge, UK

Supervisors
Dr. Alex W. Chin
L'institut des Nanosciences de Paris,
Sorbonne Université & CNRS
Paris, France

Dr. Akshay Rao
Cavendish Laboratory
University of Cambridge
Cambridge, UK

ISSN 2190-5053 ISSN 2190-5061 (electronic)
Springer Theses
ISBN 978-3-030-85453-9 ISBN 978-3-030-85454-6 (eBook)
https://doi.org/10.1007/978-3-030-85454-6

© The Editor(s) (if applicable) and The Author(s), under exclusive license to Springer Nature Switzerland AG 2021
This work is subject to copyright. All rights are solely and exclusively licensed by the Publisher, whether the whole or part of the material is concerned, specifically the rights of translation, reprinting, reuse of illustrations, recitation, broadcasting, reproduction on microfilms or in any other physical way, and transmission or information storage and retrieval, electronic adaptation, computer software, or by similar or dissimilar methodology now known or hereafter developed.
The use of general descriptive names, registered names, trademarks, service marks, etc. in this publication does not imply, even in the absence of a specific statement, that such names are exempt from the relevant protective laws and regulations and therefore free for general use.
The publisher, the authors and the editors are safe to assume that the advice and information in this book are believed to be true and accurate at the date of publication. Neither the publisher nor the authors or the editors give a warranty, expressed or implied, with respect to the material contained herein or for any errors or omissions that may have been made. The publisher remains neutral with regard to jurisdictional claims in published maps and institutional affiliations.

This Springer imprint is published by the registered company Springer Nature Switzerland AG
The registered company address is: Gewerbestrasse 11, 6330 Cham, Switzerland

Στην Μητέρα μου Κορνελία, για όλα τα βιβλία που μου αγόρασε όταν ήμουν παιδί.

Στον Πατέρα μου Γιάννη, που μου έμαθε να μην τα παρατάω ποτέ.

Στη μνήμη που Δρ. Παναγιώτη Κομνηνού, για όλες τις συζητήσεις μας σχετικά με τη Φυσική και τη ζωή στο εξωτερικό.

Supervisor's Foreword

Organic semiconductors hold tremendous potential as the active layers in the next generation of optoelectronic materials built from cheap, non-toxic and earth-abundant elements. Their versatile and tuneable optoelectronic properties have already been commercialised in flexible display technologies, i.e. organic light-emitting diodes (OLEDs), and have also been deployed in organic photovoltaic (OPV) modules and molecular electronics [1]. However, unlike the crystalline inorganic materials (Si, GaAs, etc.) that are more commonly used, today, for these applications, the fundamental charge excitations in organics interact very strongly with each other, with light and also the mechanical fluctuations (phonons) of their typically disordered condensed phases. Unlocking the potential of these materials through rational optimisation thus requires detailed understanding of their complex, non-perturbative electronic physics, and this already sets a daunting many-body problem in front of those wishing to realise this goal [2]. Moreover, as ultrafast optical experiments increasingly reveal the detailed dynamics of charges and excitons over femto-nanosecond timescales, it is becoming increasingly appreciated that an entire static picture of eigenstates and energy levels of various material blends is not enough: dissipative non-equilibrium dynamics, including even transient coherence effects, may well prove to be just as important as thermodynamics in determining their functional capabilities and efficiencies [3].

In order to make any progress, this situation essentially invites the researcher to consider the dismal prospect of solving the full time-dependent Schrödinger equation for strongly interacting charges, photons, phonons and spins. However, the situation is far from hopeless, and in this doctoral thesis, Dr. Alvertis attacks the problem from multiple angles, probing each with different state-of-the-art computational approaches whose results, taken together, provide a rich multidimensional insight into the structure and dynamics of excitations in molecular semiconductors.

After a compelling theoretical overview of the optoelectronic physics of molecular solids, Chap. 5 deals with the ultra-strong coupling between photons and excitons in organic nanowires of defect-free polydiacetylene (PDA) chains. Using ab initio electronic structure theory (GW+BSE), Dr. Alvertis shows how the very large dipole strength of the PDA exciton leads to the formation of hybrid excitations—superpositions of light and matter known as polaritons—that dominate the reflectivity of PDA surfaces. The 'photonic' properties of the states could allow them to carry excitonic energy over much larger distances than exciton diffusion allows, but while theory predicts very stable *propagating* polariton states, experiments have difficulty detecting them. Often, conformational disorder is invoked to explain the differences between theory and experiment, but here—in pristine PDA—Dr. Alvertis was able to show that strong phonon interactions are the determining factor that gives light–matter coupling the characteristics that were measured by our collaborators in Cambridge [4]. Indeed, the role of vibrations in modifying optoelectronic properties and processes is, as we shall see, a central and continuous theme throughout this thesis.

Chapter 6 picks this up in a tour de force of computational electronic structure theory, in which density functional theory, many-body perturbation theory (GW+BSE) and non-perturbative determination of exciton–phonon couplings are used to predict the excitonic properties of polyacene films from naphthalene to pentacene. Not only does this approach predict absorption spectra that are extremely close to the experiments, the inclusion of phonons also allows the lattice temperature dependence of the excitons to be extracted. Here, we encounter something new. Comparison of the singlet and triplet across the acene series allowed the experimental trends to be rationalised on the basis of the microscopic exciton wave functions themselves: as the electron–hole separation of the exciton becomes larger, the exciton primarily couples to longer-wavelength, lower-frequency phonons that are increasingly thermally activated. Triplet excitons are very tightly bound in all acenes and consequently couple to high-frequency modes with essentially no temperature dependence. In contrast, the singlet excitons show increasing electron–hole separation as the acene size increases, leading to a stronger coupling to thermally activated phonons. These results—backed by experimental measurements—are all suggestive of a tuneable link between electron–hole delocalisation, phonons and operating temperature, which could perhaps be exploited for tuning excitonic transport properties [5].

When referring to 'phonons', as above, the reader may have in mind the long-wavelength vibrational modes associated with simple inorganic crystals. These exist in molecular solids, also, but perhaps even more important are the intramolecular vibrations of the individual molecules that make up a non-covalent organic crystal. These vibrating structural deformations can be very strongly coupled to electronic excitations, especially—as discussed above—when the electron and hole are co-localised, as in a Frenkel exciton. Chapter 7 takes a deep dive into one of these elementary cells and focuses on the non-perturbative real-time dynamics of photoexcitations in a molecular dimer. After deriving an *ab initio* linear vibronic model for this system, the problem turns to how to simulate a quantum system consisting of a few electronic levels coupled to several hundred, fully quantised vibration modes. In the example presented in this chapter, this amounts to, not inconsiderable, 10^{156} states to handle! This problem is overcome with the use of tensor network, machine learning and entanglement renormalisation techniques which permit a systematic and powerful truncation of the problem to a tractable number of parameters [6]. Applying these techniques to the case of di-tetracene, Dr. Alvertis unveils how the excess energy following photon excitation is dissipated by vibrations and can be preferentially delivered into modes that promote non-equilibrium processes such as 'hot' charge separation [7]. Chapter 8 then uses the same powerful set of techniques to predict the real-time vibronic dynamics of a di-pentacene dimer, as it undergoes the process of singlet fission. Here, the many-body nature of the simulations allows the behaviour of the vibrational 'environment' to be extracted and visualised, providing a novel—fully quantum—'movie' of the fission process in terms of the motions of the molecule atomistic structure. This essential structural (vibrational) record of the fission process was then spectacularly verified by and presented alongside the results of a femtosecond, time-resolved pump–dump–probe experiment [8].

Finally, having covered the impact of strong coupling to light, delocalised phonons and intramolecular vibrations, it may seem as if we have exhausted all of the degrees of freedom in the problem. Dr. Alvertis then turns to look at the role of dynamical dielectric environments on organic photophysics, in the context of molecules in solution. Again, looking at the molecular dimer of tetracene, an incredibly rich physics, including a fundamental change in the mechanism for singlet fission as a function of solvent polarity, is rationalised through a combination of first-principles modelling, molecular dynamics, and an intuitive effective model that accounts for the key, transient symmetry breaking caused by fluctuations of the external solvent's polarisation field [9].

Seeing the sheer diversity of computational techniques and problems that Dr. Alvertis has tackled in this thesis could well provoke vertigo on a first reading of the content page, and I therefore caution the reader to take a seat before doing so. It is an absolute pleasure and honour for me to write the preface for this expansive, thought-provoking and scholarly work, just as much it was to collaborate with Dr. Alvertis over the years in which it was produced. Not only do many of the hard-won results in this monograph deepen our understanding of elementary processes in organic semiconductors, several of the observations to be found here also push in bold new directions along which only new computational techniques will be able to carry us. Even at this early stage of his career, I should not be surprised to see Dr. Alvertis in the vanguard of this movement, where combining the theory of atomistic and crystal electronic structure with real-time dynamics will reveal fascinating new physics and materials that could play a decisive role in the transition to a sustainable world.

Paris, France
July 2021

Dr. Alex W. Chin

References

1. Bronstein H, Nielsen CB, Schroeder BC, McCulloch I (2020) The role of chemical design in the performance of organic semiconductors. Nature Rev Chem 4:66
2. Barford W (2005) Electronic and optical properties of conjugated polymers. International series of monographs on physics (Oxford, England); 129. Clarendon Press, Oxford

3. Brédas JL, Sargent EH, Scholes GD (2016) Photovoltaic concepts inspired by coherence effects in photosynthetic systems. Nature Mater 16:35
4. Alvertis AM, Pandya R, Quarti C et al (2020) First principles modeling of exciton-polaritons in polydiacetylene chains. J Chemical Physics 153:084103
5. Alvertis AM, Pandya R, Muscarella LA et al (2020) Impact of exciton delocalization on exciton-vibration interactions in organic semiconductors. Phys Rev B 102:081122(R)
6. Schröder FAYN, Turban DHP, Musser AJ et al (2019) Tensor network simulation of multi-environmental open quantum dynamics via machine learning and entanglement renormalisation. Nature Commun 10:1062
7. Alvertis AM, Schröder FAYN, Chin AW (2019) Non-equilibrium relaxation of hot states in organic semiconductors: impact of mode-selective excitation on charge transfer. J Chem Phys 151:084104
8. Schnedermann C, Alvertis AM, Wende T et al (2019) A molecular movie of ultrafast singlet fission. Nature Commun 10:4207
9. Alvertis AM, Lukman S, Hele TJH et al (2019) Switching between coherent and incoherent singlet fission via solvent-induced symmetry breaking. J Am Chem Soc 141:17558

Abstract

Organic semiconductors are materials that are promising for novel optoelectronic applications, such as more efficient solar cells and LEDs. The optoelectronic response of these materials is dominated by bound electron–hole pairs called excitons, which are often strongly affected by hundreds of possible molecular vibrations. Although quantum theory contains all the ingredients to describe these complex phenomena, in practice it is only possible to solve the corresponding equations in small systems with few vibrations. As a result, it has been common to assume weak exciton-vibration interactions and to employ perturbative approaches. Similarly, exciton-photon interactions have almost universally been treated in the so-called weak coupling regime. However, in recent years it has become increasingly clear that these approximations can break down in organic semiconductors, placing an important roadblock towards the novel energy-harvesting technologies that could be based on these materials.

In this thesis, we address this issue by developing methods to treat exciton-photon and exciton-vibration interactions, without relying on any approximation regarding their magnitude. We propose a first-principles description of hybrid exciton–light (polariton) states that result from strong exciton-photon interactions. We discuss a method to treat strong exciton-vibration interactions, showing that the spatial extent of exciton states controls their magnitude. Subsequently, we present a beyond Born–Oppenheimer method based on tensor networks to study real-time exciton dynamics. By using these methods, we show how selective excitation of vibrational modes can enhance charge transfer. Moreover, through rigorous comparison to experiments, we highlight that tensor network methods are highly accurate, and we generate a 'movie' of the photophysical process of singlet fission, which occurs during early light-harvesting by organic molecules and has the potential to increase solar cell efficiencies. Finally, we construct a singlet fission model including the effects of excess energy, vibrations and the solvent of molecules concurrently, demonstrating that the fission mechanism can be qualitatively changed in a controlled manner, allowing for its acceleration by an order of magnitude.

Acknowledgements

My trip towards the Ph.D. degree has been a very exciting and pleasant one, and this is largely due to the people that I interacted with during these years of my life.

I am very grateful to my supervisor Dr. Alex Chin, who always inspired me to think outside the box and allowed me freedom to pursue any research direction that I found interesting. Despite him having to move to Paris very early in my Ph.D., he was always present and supportive throughout this academic journey of mine. I am also very thankful to Alex for all the fun times in Paris and elsewhere, Ghent, Mons, Warwick and more that I might forget.

I owe a great deal of thanks to Dr. Akshay Rao, who was happy to step up to the role of my official supervisor when Alex moved to Paris. Not only did he make me feel very welcome in his group and gave me a place to thrive, but he was also always very happy to hear my ideas and consider joint experimental–theoretical projects that led to a lot of the work presented in this thesis and beyond. When I was working away from Cambridge, he would always check whether all was well. I also thank Akshay for all the valuable career advice.

My Ph.D. experience would not have been the same had I not met Dr. Bartomeu Monserrat towards the end of my first year. Tomeu was very happy to collaborate with me and always very patient with teaching me about his methods and their physical background. He was happy for me to lead our projects in the directions that I saw fit and inspired in me a great respect for his rigorous way of examining nature. I am thankful for all this, as well as all our chats beyond science, the fun Christmas party, the ice cream from Jack's we had at Christ's piece and the walk through the Trinity's Fellows' Garden.

Already since the time of studying for my M.Phil. degree at Cambridge, my interactions with Prof. Sir Richard Friend had a great impact on me. Despite being very busy, Richard was always keen to discuss interesting science and taught me ways of developing a powerful intuition about problems before delving into computation. Perhaps most importantly, he taught me that one of the greatest virtues of a good scientist is to be a nice person.

I owe a great deal of gratitude to the director of my CDT programme Dr. Nikos Nikiforakis who always offered guidance and was ready to answer any silly question.

I thank Nikos for all his work towards making the CDT a great experience and making sure that all of the students, including myself, were doing well.

Being a member of the community of the Winton Programme for the Physics of Sustainability undoubtedly had a very positive impact on my Ph.D. experience. Thanks to Winton, I first travelled to Berkeley, spoke to several great scientists and eventually returned there for a two-month research stay at the group of Professor Jeffrey Neaton. Special thanks go to Winton Manager Dr. Nalin Patel, for our numerous conversations, for all the advice and for always being willing to help in any way possible.

I am thankful to Dr. Timothy Hele, who mentored me throughout several stages of my Ph.D. and was always a great source of ideas. Tim reminded me of the beauty and strength of a good pen and paper calculation. I am also thankful to him for all our lunches, tea breaks and zoom calls, and for all the advice.

I am indebted to Dr. Florian Schröder for teaching me all about tensor networks during my M.Phil. and the first weeks of my Ph.D., patiently walking me through his code, despite being extremely busy with finalising his thesis. A big part of the work presented in this thesis was only made possible due to Florian's help.

I am grateful to Dr. David Turban for explaining several aspects of the electronic structure of DP-Mes during my early Cambridge days, helping me to figure out practical issues with my calculations and also sharing his Python scripts that made my work much simpler.

I would also like to thank Dr. Christoph Schnedermann, for helping me understand how theory and experiment can work hand-in-hand and for teaching me the value of a good graph. I am also thankful to Chris for forcing me to take coffee breaks and not letting me work non-stop and despair over my laptop.

Many thanks to Dr. Andrew Musser who offered me the possibility to work on the interesting problems that he first studied experimentally and for all the interesting discussions around them. I thank him for being patient with the first-year inexperienced theoretician (myself) who was trying to interpret his complex data and finally for the hospitality when I visited him at Cornell University.

I would like to thank Dr. David Beljonne who hosted me in his group in Mons for two and a half months and taught me so much about hands-on electronic structure calculations, always being there to answer any silly question. I am also indebted to Claudio Quarti, a postdoc in Mons at the time, for helping me with a lot of the practical issues I faced with these calculations, but most importantly for being a good friend and making my stay a pleasant one. With this, I also thank the rest of the team that we always had drinks and dinner with: Valentin, Elliot, Manuel and the others.

I am grateful to Dr. Thierry Barisien and Dr. Laurent Legrand for being great hosts during all of my visits in Paris. I have several long, science-themed and very fun lunches to remember. Further, I thank Thierry and Laurent for our long scientific discussions in the Jussieu seminar rooms, where we would stay for hours until we made sense of the latest puzzling piece of physics around polydiacetylene.

I thank professor Jeffrey Neaton for hosting me in his group at UC Berkeley for two months, in what proved to be a fascinating experience. I am grateful to him for always offering to help, without me having to ask, and for making me feel so welcome. I am also grateful to a lot of other people at Berkeley, who all contributed in a scientifically and socially very pleasant interaction: Jonah Haber, Felipe Jornada, Marina Filip, Liz Peterson, Samia Hamed and more.

There are a lot of people I could name from the Rao group that I owe thanks to. Doing a Ph.D. in the group was always great fun, and being the only theoretician among so many experimentalists had its challenges. But I never felt isolated and people were keen to discuss ideas that could lead to interesting new research avenues. And let us not forget our countless lunches in the Maxwell building, Christmas dinners, playing cards against humanity and so many other beautiful memories that I will always cherish. It is difficult to pick a few people from the group to mention. Nevertheless, here I will attempt to do this, and hopefully anyone I leave out will forgive me and also accept my thanks.

I am thankful to Dr. Raj Pandya for all the fun times. I thank Raj for agreeing to run the experiments that I asked him for and working tirelessly on them. It was always a pleasure discussing science and seeing the energy and enthusiasm with which he would run around the building. It was great fun being together in Paris a few times, as well as being in Athens in the most peculiar August of 2020 over the course of a pandemic.

Special thanks go to my longest-standing office mates in Cambridge: Dr. Jooyoung Sung and Dr. Sachin Dev Verma. Our being together in the office sparked several spontaneous scientific discussions and was always great fun. I thank them for putting up with me for the best part of three years. I am also grateful to Dr. Victor Gray and Dr. Zhaojun Li who joined the office later and were always very nice to me. Although experimental chemists and theoretical physicists usually do not mix well, we managed to make it work!

I am thankful to Dr. Ture Hinrichsen for all the fun puns and for the great trip in Tuscany, also together with Jooyoung. I even got to rub sand on Ture's face in Bagno Vignoni! I will never forget the gorgeous places that we all saw together, the great food we had and all our laughing.

I am grateful to Arjun Ashoka for all the ice cream we had and for all the questions that he insisted on asking me, which helped me realise whether I really understood something. I thank Arjun for his contaminating enthusiasm about science and for being a good friend and collaborator.

I am thankful to Sasha for all the fun chats and to Hope and Arya for always making sure that everyone in the group was feeling welcome. I am also thankful to Arya and Hope for a very fun day in Denver, when we almost got stuck during the outbreak of the pandemic.

In my four years in Cambridge, I made many great friends, many of whom I hope will remain an important part of my life. Dr. Eleftherios Ioannou, my dear friend Lefteris, comes to mind, who made sure that my transition to Cambridge was a smooth one, who joined me for all my late lunches, and with whom we shared so many stories and coffee breaks. I thank my good friends Vasilis and Stratis for all the incredibly fun times in Cambridge and Athens, for hosting me in their house countless times, for sharing their food which was always tasty, for the music evenings, for having drinks at Darbar and so much more that I could never put on paper. I thank

Iasonas and Georgia that together with Stratis formed my St Edmund's crew of my last year, making me to always hang around college and looking forward to our dinners together and to our chats in the CR.

Almost from my first day in Cambridge, and up to the last as a Ph.D. student, I stayed in Link House. It was amazing how many good friends from all over the world I made in this international community. I am thankful to the wardens, Gary and Anne, for making sure that the atmosphere was always so great and for always being there when I needed support in any way possible. There is a lot I could say about my friends from the house, but a big thanks to Daniele and Jee-Yeon, Diego, Elena, Toshi and Chiara will have to suffice for now. Even lockdown was a nice time in Link House, for which special thanks go to Beatrice for all the fun things we did and for teaching me so much about movies.

I would never have managed to achieve the things I did had it not been for my family and friends in Greece. I cannot imagine that one could wish for better and more supportive parents than my mother Cornelia and my father Yannis. Without them, and without my grandparents Antonis and Vasiliki, I would not be the person I am today. I owe eternal gratitude to our family friend Dr. Panagiotis Komninos, who so tragically passed away. It was largely thanks to our countless discussions on physics and what it is like to be an academic abroad that I was able to obtain a better understanding of the possible paths I could follow. It was also always such good fun to hear his stories about the Greek community in Munich. Finally, I owe the deepest thanks to my aunts Zeta and Mary for always being there in the difficult times and my aunt Monica and cousin Marina for their unconditional love.

Contents

Chapter 1
Introduction

Over millions of years of evolution, nature has perfected the process of photosynthesis in plants. It is only natural to wonder whether humans could mimic photosynthesis using synthetic organic molecules, in order to harness sunlight for society's benefit. Solar energy provides a sustainable alternative to carbon fuels and can sustain future life, making this is a much desirable goal.

One class of organic synthetic molecules that have the potential to achieve efficient light-harvesting are organic semiconductors. These materials have attracted a lot of attention during the past few years due to their promise for novel optoelectronic technologies, such as organic photovoltaics and LEDs. It was however key discoveries from a few decades ago that set the foundations, with key moments including the demonstration of the ability of organic semiconductors to exhibit large conductivity [1] and to emit light upon the application of an external voltage [2].

Following photon incidence on a molecule, its energy is carried by bound pairs of electrons and holes, i.e. the charge carriers of a molecule. These bound electron-hole pairs are referred to as excitons and are the carriers of energy; it is their generation, transfer across interfaces and transport over long distances that is of paramount importance to efficient light-harvesting and solar energy storage. Hence optimising light-harvesting in molecular systems requires maximising exciton-photon interactions and optimising excitonic motion.

At a first level this can be treated as a purely electronic problem: the absorption of a photon generates excitons (this is referred to as photoexcitation), which can be transferred to other molecules and transported across certain distances. The efficiency of these processes will depend on quantities such as the overlap of the electronic wavefunctions of the different components of the studied system, their relative energies, etc.... While this purely electronic picture for light-harvesting processes in organic semiconductors has led to several insights, in recent years it has become increasingly clear that reality requires a more comprehensive understanding. Simultaneous to the motion of excitons, a molecule vibrates in hundreds of possible ways, according to its normal modes of vibration. While in several material classes these vibrations may be regarded as a small perturbation to the motion of charges/excitons, this is not the case in molecular systems, where exciton-vibration interactions can be

© The Author(s), under exclusive license to Springer Nature Switzerland AG 2021

A. M. Alvertis, *On Exciton–Vibration and Exciton–Photon Interactions in Organic Semiconductors*, Springer Theses,
https://doi.org/10.1007/978-3-030-85454-6_1

extremely strong. Computing these strong interactions is in itself challenging, and also leads to complications when it comes to studying the time-evolution of these systems; vibrations can induce non-adiabatic transitions between electronic states, leading to a breakdown of the commonly employed Born-Oppenheimer approximation [3]. Moreover, exciton-photon interactions can also become very strong in some molecular systems, leading to deviations from the so-called weak coupling limit, within which absorption is considered a 'one-off' event, during which the energy of a photon is transferred to an exciton instantaneously.

This thesis is concerned with providing a deeper understanding of the physics of strong exciton-photon and exciton-vibration interactions that are present in molecular systems, hence providing a stepping-stone towards achieving efficient man-made light-harvesting systems. The first part provides the background that will allow the reader to better follow the results. In particular, Chap. 2 provides an overview of organic semiconductors and some of their unique physical properties. Chapter 3 discusses the formalism used to study the time evolution of a system within quantum mechanics, where we also discuss shortcomings of commonly employed methods when it comes to studying systems interacting strongly with an environment e.g. of vibrations. Chapter 4 is concerned with the modelling of electronic and vibrational structure in equilibrium, outlining the techniques used throughout this thesis in order to obtain several properties of electrons and holes, excitons and the vibrations of organic semiconductors.

The second part of the thesis presents a series of novel contributions towards the goal that has been outlined above. In Chap. 5 we explore strong exciton-photon interactions, which are of relevance at the earliest stages following photoexcitation. We outline ways of tuning the magnitude of these interactions by developing a first-principles methodology for their calculation.

In Chap. 6 we study the strong exciton-vibration interactions that are present in the acene series of organic semiconductors. When it comes to the problem of light-absorption, these are mostly relevant at later timescales compared to exciton-photon interactions. We show that the spatial extent of excitons largely determines the magnitude of their interaction with molecular and crystal vibrations. We thus present a theory that allows us to uncover the full microscopic mechanism behind exciton temperature- and pressure-dependence, rigorously comparing against experiment in both cases.

In Chaps. 7–9 we proceed to study the motion of excitons and their interaction with vibrations in the time domain, following the excitation of a system by a photon. We study so-called ultrafast processes occurring in the femto- to picosecond timescale following photoexcitation, discussing the microscopic exciton-vibration interactions that occur and how these enable the optimisation of the efficiency of the various processes.

In Chap. 7 we study the process of endothermic charge transfer in a covalent tetracene dimer in timescales of up to four picoseconds. By using and expanding upon a highly accurate, non-perturbative and beyond Born-Oppenheimer method based on tensor networks, we show that the activation of molecular vibrations through excess energy allows for a more efficient transition from a spatially localised to a charge transfer exciton. Molecular vibrations of different energies have different effects on charge transfer, allowing us to discuss the potential of vibrational mode-selective excitation as a means of enhancing the efficiency of ultrafast processes.

In Chap. 8 we proceed to apply the tensor network methodology discussed in Chap. 7 to the process of ultrafast singlet exciton fission, which occurs within less than a picosecond and during which a spin-singlet exciton decays towards two spin-triplet excitons. By comparing our theoretical results to experiment, we are able to showcase the accuracy of our tensor network methods and to reveal the full microscopic mechanism of singlet fission in a covalent pentacene dimer, creating a 'movie' of the molecular motion. We show that molecular vibrations of different nature coordinate in a precise way in order to enable this process.

In Chap. 9 we combine the insights from Chaps. 7 and 8 in order to construct a simplified model for studying the process of endothermic singlet exciton fission at significantly longer timescales of up to hundreds of picoseconds. Our model includes the effects of molecular vibrations, excess energy and the dielectric environment of the molecular system. We show how a deep understanding of the relevant photophysics can be used to find the subtle balance between these different factors that can lead to important efficiency gains.

Finally, we would like to highlight that the presented thesis contains various kinds of computational methods, which in some cases are novel, in some they constitute extensions to pre-existing methods, and in some other cases they are combinations of methods that had never before been combined. The applicability of these methods goes beyond the photophysics of organic semiconductors, and can be used to study the complicated quantum mechanical problem of excited states interacting with a vibrational environment in numerous material systems.

References

1. Chiang CK, Fincher CR, Park YW, Heeger AJ, Shirakawa H, Louis EJ, Gau SC, MacDiarmid AG (1977) Electrical conductivity in doped polyacetylene. Phys Rev Lett 39:1098
2. Friend RH, Gymer RW, Holmes AB, Burroughes JW, Marks RN, Taliani C, Bradley DDC, Dos Santos DA, Bredas J-C, Logdlund M, Salaneck WR (1999) Electroluminescence in conjugated polymers. Nature 397:121
3. Born M, Oppenheimer JR (1927) Zur Quantentheorie der Molekeln. Annalen der Physik 84:457

Part I
Background

Chapter 2
Organic Semiconductors and Their Properties

Carbon atoms form the basis of organic molecules and polymers, which are the building blocks of organic semiconducting materials. There are millions of possible organic structures, which allow one to imagine countless ways of designing materials with the desired properties. The small atomic number of carbon, as well as hydrogen, nitrogen and oxygen, which are the other elements often participating in organic compounds, gives organic semiconductors some of their unique properties, compared to their inorganic counterparts such as silicon. In this chapter, we explore the key characteristics of organic semiconducting materials, which has been the subject of classic textbooks such as reference [1]. Here we start from the basic properties of carbon atoms, and discuss the role of low screening due to the small number of electrons, the distinction between molecular and crystal vibrations, and finally we draw a comparison to inorganic semiconductors.

2.1 The Chemistry of Carbon and π-conjugation

We start by considering the ground electronic state of a carbon atom. The six electrons are distributed among the lowest energy eigenstates, that is the atomic orbitals of carbon, according to the rules that Friedrich Hund formulated in 1925 [2]: $1s^2 2s^2 2p^2$, as schematically represented in Fig. 2.1a. There is a total of four valence electrons in the outermost $2s$ and $2p_x$, $2p_y$, $2p_z$ atomic orbitals, however only two of them are unpaired and available for the formation of a chemical bond. Nevertheless, carbon atoms can form substances such as methane (CH_4), with up to four covalent bonds. Furthermore, the chemical bonds of methane are known to be equivalent, i.e. there is no indication that some of the bonds are formed between the carbon $2s$ orbital and the hydrogen $1s$ orbital, while others between the carbon $2p$ and hydrogen $1s$ orbitals. In order to explain these apparent contradictions, Linus Pauling introduced a theory of chemical bonds in 1931, based on the hybridisation of atomic orbitals [3]. This so-called valence bond theory suggests that one of the $2s$ electrons is promoted into the empty $2p$ orbital, and new eigenstates are formed by mixing the s and p orbitals.

© The Author(s), under exclusive license to Springer Nature Switzerland AG 2021

A. M. Alvertis, *On Exciton–Vibration and Exciton–Photon Interactions in Organic Semiconductors*, Springer Theses,
https://doi.org/10.1007/978-3-030-85454-6_2

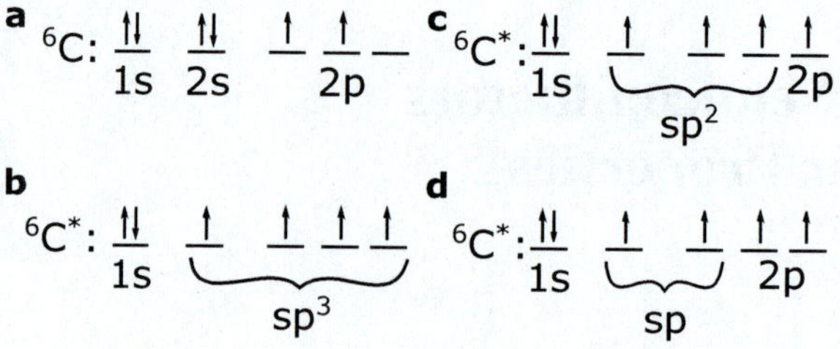

Fig. 2.1 The electronic configuration of a carbon atom in its ground state, as well as in hybridised states encountered in chemical bonding

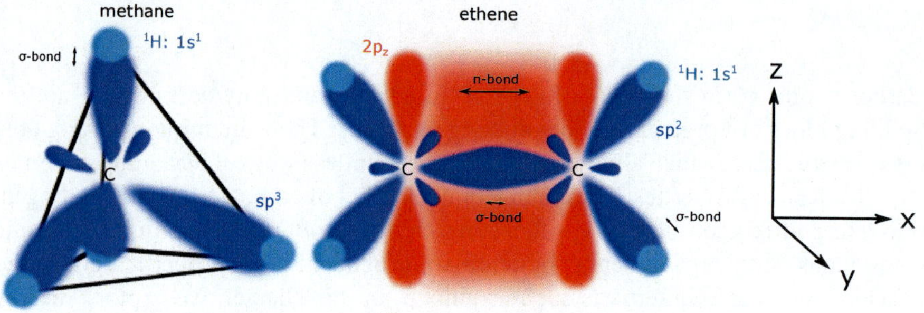

Fig. 2.2 A schematic representation of the chemical bonding in methane and ethene, according to valence bond theory

These hybrid orbitals are referred to as sp^3, are degenerate in energy and arrange themselves in a tetrahedron around the carbon atom, as visualised in Fig. 2.2 (left). The orbital configuration of an sp^3 hybridised carbon atom is given in Fig. 2.1b; the star symbol in ${}^6C^*$ indicates that the carbon atom is in an excited configuration, however the energetic stabilisation provided by the chemical bonding outweighs this increase in energy, making bond formation energetically favourable. Each of the carbon hybrid orbitals interacts with the s orbital of a hydrogen atom, forming strong σ bonds as a consequence of the significant overlap of their electronic densities.

However, it is not always the case that carbon atoms form four equivalent chemical bonds. For example, in an ethene molecule (C_2H_4), the two carbon atoms form a double bond, and one of the two is weaker than the other. In practice, this means that it is relatively easy to break the double bond and form two single bonds, which gives ethene some of its interesting chemical reactivity properties. The concept of hybridisation of atomic orbitals within valence bond theory may be used to explain the properties of ethene. In this case, only two of the valence p orbitals of carbon mix with the $2s$ orbital, forming what is called sp^2 hybrids (Fig. 2.1c). These participate in strong σ bonds with the two hydrogen atoms bonded to each carbon, and in a single 'face-to-face' σ bond between the two carbon atoms. The remaining unhybridised $2p$ orbitals of each carbon lie parallel to each other, showing a small yet finite overlap of their electronic density, which leads to a weak π bond between the two carbon atoms. Chemical bonding in ethene is schematically represented in Fig. 2.2 (right).

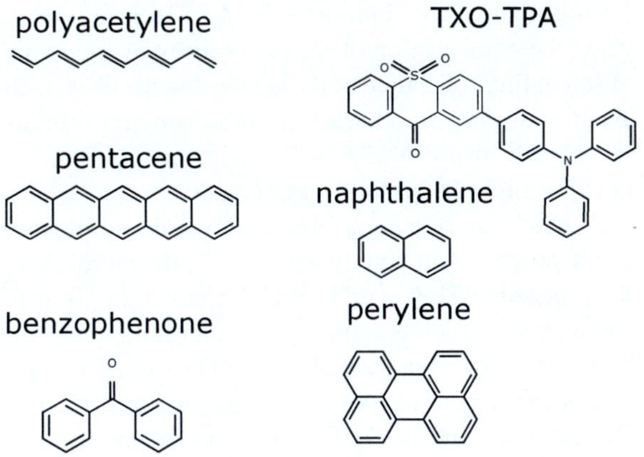

Fig. 2.3 Some example conjugated organic molecules, which exhibit semiconducting properties

In a similar fashion, in molecules such as ethyne (C_2H_2), where carbon atoms form a triple bond, only a single p orbital mixes with the s orbital to form two sp hybrids. The distribution of electrons in the carbon atomic orbitals in this case is shown in Fig. 2.1d. If we define the x axis of the system to lie in the direction of the carbon-carbon σ bond, then the unhybridised $2p$ orbitals lying in the y and z directions participate in π bonds.

The chemical bonding properties of carbon are what give this element its unique properties and the ability to create millions of possible structures. Carbon atoms can bond to each other with single, double or triple bonds, forming linear chains, branches of those, cyclic compounds, in what is a set of nearly endless possibilities. An important subset of organic molecules is the so-called conjugated structures. Some of them are shown in Fig. 2.3. Due to the weak bonding and delocalised character of the π electrons associated with a multiple bond, this alternation in the structure of conjugated materials leads to π electrons that are delocalised over the entire molecule, and do not strictly belong to a single pair of atoms, as described within valence bond theory. It is therefore natural to describe these delocalised electrons as belonging to electronic states associated with the molecule as a whole, rather than atomic orbitals. These states are referred to as *molecular orbitals* and can be constructed as linear combinations of atomic orbitals (LCAO). The theory of molecular orbitals was first introduced by Hund, Mulliken and Lennard-Jones between 1927 and 1929 [4–8]. In 1931, Hückel applied this method in the context of conjugated systems, forming linear combinations of the electrons originating from unhybridised p orbitals. The μth molecular orbital describing a delocalised π-electron state is then written as [9]:

$$\pi^{\mu} = \sum_{i=1}^{N} c_i^{\mu} p_z^i, \tag{2.1}$$

where N the number of carbon atoms participating in the conjugated structure. The coefficients c_i^{μ} are determined by solving a generalised eigenvalue problem, for a

Hamiltonian which ignores electron–electron interactions. In the example of benzene, the combination of the six p_z orbitals gives six π molecular orbitals, as visualised in Fig. 2.4. Three of these have lower energies than a system of six independent p_z orbitals, and are characterised as bonding orbitals. These are the orbitals populated by electrons. Additionally, there are three anti-bonding molecular orbitals, with energies higher than the p_z system.

Hückel theory is very useful for obtaining physical intuition regarding the electronic properties of conjugated organic structures, and theoretical chemists have used it for decades. However, in this thesis we do not use Hückel to perform any actual calculations, as it has some significant limitations, which can nowadays be overcome by using more advanced schemes. For example, within Hückel theory, σ electrons do not participate in forming the molecular orbitals, instead forming a background electronic density. In 1963, Roald Hoffmann presented an extended Hückel theory, also incorporating the effect of σ electrons, and achieving improved accuracy for the electronic properties of molecules [10]. Furthermore, the repulsive Coulomb interaction between p_z electrons is ignored. In Sect. 4.3 we discuss the so-called PPP theory introduced by Parr, Pople and Pariser in [11, 12], which is used within this thesis. This approach is still based on a model description of π electrons, however it does incorporate electron correlation, which is necessary in order to achieve quantitative accuracy for optical properties of some of our studied systems. In several cases however, we need to resort to fully first-principles calculations that do not depend on approximations regarding the σ electrons, the form of the Coulomb interaction, or LCAO. These include density functional theory and Green's function based methods discussed in detail in Chap. 4.

The highly delocalised character of π electrons in conjugated materials led to the assumption that these materials could exhibit metallic conductivity, hence sparking intense research efforts in that direction. It was however subsequently realised that conjugated organic structures are not in fact conducting, since they exhibit a relatively large gap (often more than 2 eV) between their highest occupied molecular orbital (HOMO) and the lowest unoccupied molecular orbital (LUMO), similar to the gap between the valence and conduction band of traditional semiconductors. These materials are therefore referred to as organic semiconductors, and the structures visualised in Fig. 2.3 all belong to this category. The gap of the benzene molecule is visualised in Fig. 2.4. The large gap limits the conductivity, as at room temperature the probability of exciting carriers at energies above 2 eV is very low, similar to inorganic semiconductors. However, doping organic crystals can provide free carriers, which are highly mobile, and conductivity can increase by up to a staggering eleven orders of magnitude [13]. These organic semiconducting materials have indeed a lot of similarities to their inorganic counterparts such as silicon, however they also exhibit a lot of differences; a subject that we discuss in more detail in Sect. 2.4.

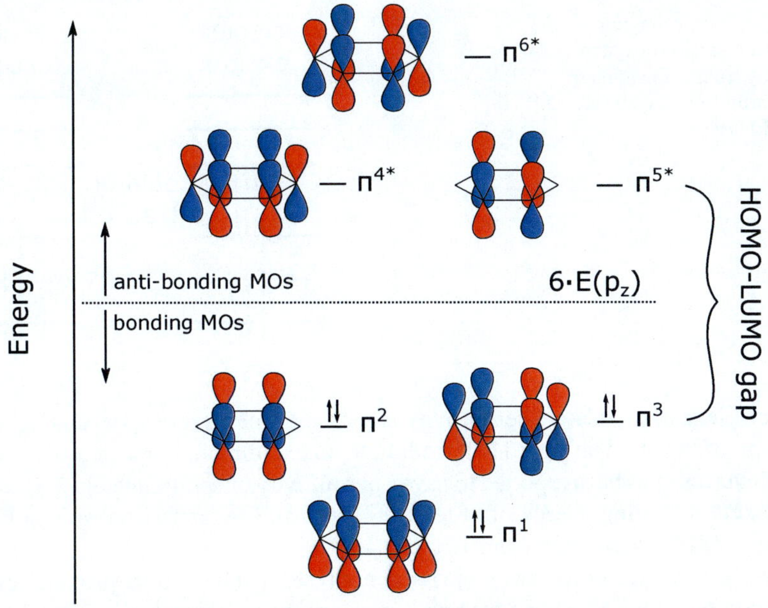

Fig. 2.4 Schematic representation of the frontier molecular orbitals of benzene, as linear combinations of p_z orbitals within Hückel theory. The blue and red colours represent positive and negative wavefunction phases respectively. In the cases that a p_z orbital is absent from a carbon atom, this means that it does not contribute towards the formation of the molecular orbital

2.2 The Exciton State

2.2.1 Formation of Bound Electron-Hole Pairs

We now consider the case of an incident photon on an organic semiconductor. Within the simplest picture, if the photon energy is larger or equal to the HOMO-LUMO gap (i.e. the fundamental gap), then there is a finite probability that it will be absorbed, promoting an electron from the HOMO to the LUMO. As a consequence, there is a surplus of negative charge $-|e|$ in the LUMO due to the presence of a single electron. Similarly, there is a a surplus of positive charge $+|e|$ in the HOMO, due to the absence of the excited electron, or equivalently, due to the presence of a hole. An example energy diagram for an exciton resulting from a HOMO-LUMO transition is visualised in Fig. 2.5. Two spin configurations are shown, and discussed below.

The excited electron in the LUMO and the hole in the HOMO interact via an attractive screened Coulomb interaction $W = -\frac{1}{\varepsilon r}$, where ε the dielectric constant of the material. Here we have adopted atomic units, which are used throughout this thesis. Compared to inorganics, organic semiconductors have low dielectric constants leading to a strong attractive electron-hole interaction in these systems. This leads to the formation of bound electron-hole pairs referred to as *excitons*. For example,

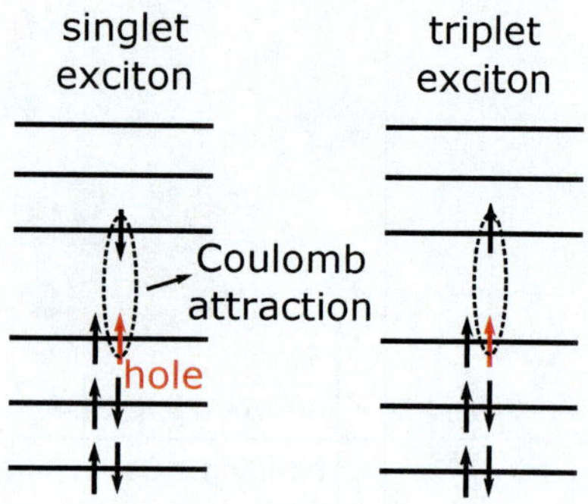

Fig. 2.5 Energy diagram of a singlet and triplet exciton resulting from a transition of an electron from the HOMO to the LUMO

polyacetylene has a relative permittivity $\varepsilon_r = \frac{\varepsilon}{\varepsilon_o}$ between 2 and 5, depending on the direction of the incident light [14], leading to an exciton binding energy of 0.15 eV [15]. Pentacene has been reported to have a permittivity of approximately 2.5, leading to an exciton binding energy of at least 0.4 eV [16]. In comparison, silicon has an exciton binding energy of a mere 15 meV [17].

More generally, exciton states ($|S\rangle$) are not restricted to frontier orbitals, and are linear combinations of transitions between occupied and unoccupied orbitals, or in other words between valence (v) and conduction (c) states:

$$|S\rangle = \sum_{v}^{\text{hole}} \sum_{c}^{\text{el}} A_{vc}^{S} \, |vc\rangle \, . \tag{2.2}$$

The coefficients A_{vc}^{S} are obtained as solutions to the so-called Bethe-Salpeter equation for exciton states, which is outlined in Sect. 4.4.3.

This picture of excitons as molecular excited states was first introduced by Frenkel in [18], and holds for a large number of organic semiconductors, both in solution and in the solid state. However, for some solid state organic semiconductors excitons can become delocalised, i.e. for a hole localised on a given molecule, the electron may be found within a radius that can include up to several monomer units. These delocalised excitons are similar to those first described by Wannier in [19] for insulating solid state systems, which are often referred to as Wannier-Mott excitons. However, the term Wannier-Mott excitons usually refers to systems with a small exciton binding energy, such as inorganic semiconductors (this is discussed in more detail in Sect. 2.4). For organics, even when excitons become delocalised, the exciton binding energy can remain substantial, as is the case with pentacene. We instead refer to these excitons as charge transfer excitons, due to the fact that the electron and hole need not be localised on the same molecule. Figure 2.6 visualises the first optically accessible exciton state of the acene series of organic semiconductors. While naphthalene hosts a localised Frenkel exciton, the charge transfer character of this bound state becomes more prominent as one moves towards pentacene, for reasons that will be discussed in detail in Chap. 6. Therefore, it becomes clear that organic semiconductors can host excitons of a varying degree of localisation.

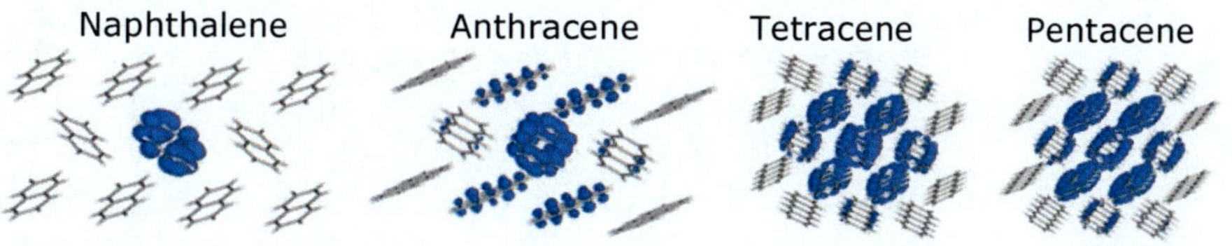

Fig. 2.6 The first bright (singlet) exciton state of the acene series of organic semiconductors. The electron density is visualised in blue, for a hole localised on the central monomer of the plotted region

2.2.2 Exciton Spin

We now turn our attention to the role of spin in the properties of excitons. An electron and a hole both have a spin of $s = 1/2$, therefore an exciton can either be in a spin singlet (S) state $S = s_{\rm el} - s_{\rm h} = 0$ ($m_S = 0$), or in a spin triplet (T) state $S = s_{\rm el} + s_{\rm h} = 1$ ($m_S = -1, 0, +1$), as visualised in Fig. 2.5 (the hole has an m_S value opposite to that of the electron state which is vacated). The corresponding spin wavefunctions are:

$$|S\rangle = \frac{1}{\sqrt{2}}(|\uparrow\rangle\,|\downarrow\rangle - |\downarrow\rangle\,|\uparrow\rangle) \tag{2.3}$$

and:

$$|T\rangle = \begin{cases} |\uparrow\rangle\,|\uparrow\rangle \\ \frac{1}{\sqrt{2}}(|\uparrow\rangle\,|\downarrow\rangle + |\downarrow\rangle\,|\uparrow\rangle) \\ |\downarrow\rangle\,|\downarrow\rangle \end{cases} \tag{2.4}$$

The singlet and triplet exciton states are non-degenerate, and the singlet has higher energy [20]:

$$E_S - E_T = 2\int d\mathbf{r}_1 d\mathbf{r}_2 \psi_v^*(\mathbf{r}_1)\psi_c^*(\mathbf{r}_2)\frac{1}{|\mathbf{r}_1 - \mathbf{r}_2|}\psi_c(\mathbf{r}_1)\psi_v(\mathbf{r}_2). \tag{2.5}$$

The derivation of the above result requires a many-body treatment of the electronic system. Such a treatment is provided by the so-called Bethe-Salpeter equation that is discussed in Sect. 4.4.3, where we also revisit the issue of the energy splitting between singlet and triplet excitons. However, an intuitive (and oversimplified) way of understanding this effect is to consider the interaction and indistinguishable character of the excited electron, and the unpaired electron that still occupies the valence state. The spin-singlet and triplet states are antisymmetric and symmetric with respect to particle exchange respectively, and the corresponding spatial parts of the wavefunction have the opposite symmetry in each case, as the product must give an antisymmetric wavefunction for fermions. Therefore, for this two-electron system the spatial part of the wavefunction is:

$$\Psi_T = \frac{1}{\sqrt{2}}[\psi_1(\mathbf{r}_1)\psi_2(\mathbf{r}_2) - \psi_1(\mathbf{r}_2)\psi_2(\mathbf{r}_1)], \tag{2.6}$$

$$\Psi_S = \frac{1}{\sqrt{2}}[\psi_1(\mathbf{r}_1)\psi_2(\mathbf{r}_2) + \psi_1(\mathbf{r}_2)\psi_2(\mathbf{r}_1)]. \tag{2.7}$$

The Hamiltonian operator for this two-electron system is: $\mathcal{H} = h_1 + h_2 + \frac{1}{r}$, where h_1, h_2 the independent electron Hamiltonians (with the respective one-electron energies $\varepsilon_1, \varepsilon_2$), and $\frac{1}{r}$ their repulsive Coulomb interaction.

The energy of the singlet is:

$$\begin{aligned} E_S &= \langle \Psi_S | \mathcal{H} | \Psi_S \rangle \\ &= \frac{1}{2}\int d\mathbf{r}_1 d\mathbf{r}_2 [\psi_1^*(\mathbf{r}_1)\psi_2^*(\mathbf{r}_2) + \psi_1^*(\mathbf{r}_2)\psi_2^*(\mathbf{r}_1)] \left(h_1 + h_2 + \frac{1}{r}\right) \\ &\quad [\psi_1(\mathbf{r}_1)\psi_2(\mathbf{r}_2) + \psi_1(\mathbf{r}_2)\psi_2(\mathbf{r}_1)] \\ &= \varepsilon_1 + \varepsilon_2 + \int d\mathbf{r}_1 d\mathbf{r}_2 \psi_1^*(\mathbf{r}_1)\psi_2^*(\mathbf{r}_2)\left(\frac{1}{r}\right)\psi_2(\mathbf{r}_1)\psi_1(\mathbf{r}_2) \\ &\quad + \int d\mathbf{r}_1 d\mathbf{r}_2 |\psi_1(\mathbf{r}_1)|^2 \left(\frac{1}{r}\right)|\psi_2(\mathbf{r}_2)|^2 \end{aligned} \tag{2.8}$$

and for the triplet:

$$\begin{aligned} E_T &= \langle \Psi_T | \mathcal{H} | \Psi_T \rangle \\ &= \frac{1}{2}\int d\mathbf{r}_1 d\mathbf{r}_2 [\psi_1^*(\mathbf{r}_1)\psi_2^*(\mathbf{r}_2) - \psi_1^*(\mathbf{r}_2)\psi_2^*(\mathbf{r}_1)] \left(h_1 + h_2 + \frac{1}{r}\right) \\ &\quad [\psi_1(\mathbf{r}_1)\psi_2(\mathbf{r}_2) - \psi_1(\mathbf{r}_2)\psi_2(\mathbf{r}_1)] \\ &= \varepsilon_1 + \varepsilon_2 - \int d\mathbf{r}_1 d\mathbf{r}_2 \psi_1^*(\mathbf{r}_1)\psi_2^*(\mathbf{r}_2)\left(\frac{1}{r}\right)\psi_2(\mathbf{r}_1)\psi_1(\mathbf{r}_2) \\ &\quad + \int d\mathbf{r}_1 d\mathbf{r}_2 |\psi_1(\mathbf{r}_1)|^2 \left(\frac{1}{r}\right)|\psi_2(\mathbf{r}_2)|^2 \end{aligned} \tag{2.9}$$

Hence their difference is exactly that of Eq. (2.5), and the integral:

$$K = \int d\mathbf{r}_1 d\mathbf{r}_2 \psi_v^*(\mathbf{r}_1)\psi_c^*(\mathbf{r}_2)\frac{1}{|\mathbf{r}_1 - \mathbf{r}_2|}\psi_c(\mathbf{r}_1)\psi_v(\mathbf{r}_2), \tag{2.10}$$

is referred to as the exchange integral. One could imagine that in the triplet state the two electrons can exchange positions due to having the same spin orientation and being indistinguishable, hence stabilising its energy relative to the singlet state. Again we emphasise that this two-electron picture offers a mere simplistic way of understanding the singlet-triplet splitting of exciton states, which needs to be considered within a many-body picture. Intuitively, the additional exchange term

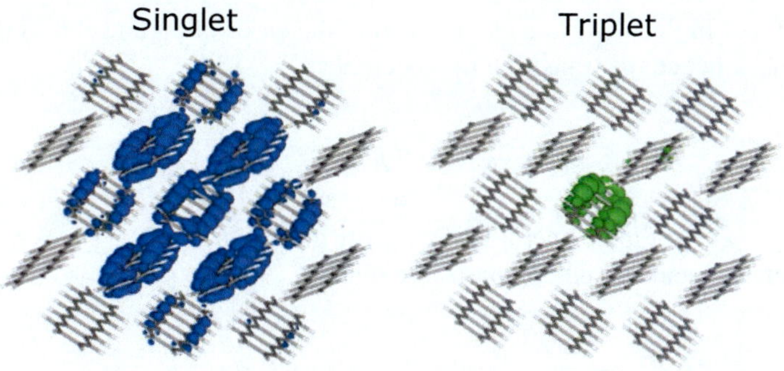

Fig. 2.7 The first singlet and triplet excitons of a molecular crystal of pentacene. The electron density is visualised in blue and green respectively, for a hole localised on the central monomer of the plotted region

destabilising the energy of the singlet can be thought of as a repulsive interaction between an electron and a hole. Consequently, singlet excitons are spatially more extended, i.e. more delocalised, than triplet excitons, as in the example of the singlet and triplet states of a pentacene crystal visualised in Fig. 2.7. We will show in Chap. 6 that the delocalisation of excitons largely determines their interactions with vibrations and their response to external stimuli such as pressure. The energy difference between the singlet and the triplet is typically 0.7–1 eV [21], a result that is critical to various ultrafast processes and optoelectronic applications based on organic semiconductors, a fact that will be discussed later on.

2.2.3 The Role of Spin for Exciton-Photon Interactions

For photoinduced processes it is important to consider the properties of the two kinds of exciton states regarding the absorption and emission of light. To do so, we will consider the coupling to the photon to be weak, making it possible to describe this interaction within time-dependent perturbation theory, and specifically using Fermi's golden rule [22]. In Chap. 5 of this thesis, we go beyond this so-called weak coupling regime and consider the physics of systems with strong exciton-photon interaction.

From Fermi's golden rule, the transition rate from an initial state $|i\rangle$ to a continuum of final states $|f\rangle$, due to a perturbation $\mathcal{H}'$, is given by:

$$k_{i\rightarrow f} = \frac{2\pi}{\hbar}|\langle i|\mathcal{H}'|f\rangle|^2\rho(E_f), \tag{2.11}$$

where $\rho(E_f)$ the density of states at the energy of the final state. In the case of photon absorption/emission, the external perturbation is an oscillating electric field of magnitude $\mathcal{E}$ in the direction $\mathbf{u}$, which interacts with the dipole generated by the transition. One can consider the electric field to be constant over the typical 1 nm length scale of molecular systems (apart from the cases of γ- and x-rays that we do not consider), hence the transition rate becomes:

$$k_{i\to f} = \frac{2\pi\mathcal{E}^2}{\hbar} | \langle i | \boldsymbol{\mu} \cdot \mathbf{u} | f \rangle |^2 \rho(E_f), \tag{2.12}$$

where $\boldsymbol{\mu} = -e\mathbf{r}$ the dipole operator. If we write the initial and final states as a product of their spin (χ) and spatial (ψ) parts, we obtain:

$$k_{i\to f} = \frac{2\pi\mathcal{E}^2}{\hbar} \left|\langle\chi_i \left|\chi_f\right\rangle \cdot \langle\psi_i \left|\mu \cdot \mathbf{u}\right| \psi_f\rangle\right|^2 \rho(E_f), \tag{2.13}$$

since the dipole operator only acts on the spatial part. Therefore, the transition rate is proportional to $\left|\langle\chi_i \left|\chi_f\right\rangle\right|^2$, which is zero for spin states of different multiplicity in the absence of spin-orbit coupling (this can be easily shown by using the spin wavefunctions of Eqs. 2.3 and 2.4). One therefore reaches the conclusion that light cannot induce a transition between a singlet and a triplet state. Since the ground state of organic semiconductors is a singlet[1] (all the electrons are paired), light absorption can only excite higher energy singlet states. This is the reason that the energy of the first bright singlet state is commonly referred to as the optical gap of a material (a quantity which is substantially smaller than the electronic/fundamental gap due to the significant exciton binding energy of organics). Similarly, a triplet exciton cannot recombine towards the ground state by emitting a photon, hence triplets are generally dark states. However, the forbidden character of singlet-triplet transitions is relaxed in the presence of spin-orbit coupling [24].

2.2.4 *Singlet Exciton Fission*

Depending on the relative energy of singlet and triplet excitons, a process referred to as singlet exciton fission can occur. During this process, a high-energy singlet exciton (typically the first excited singlet S_1) decays towards a pair of low-energy triplets (see references [25, 26] for detailed reviews on singlet fission). Hence the process is exothermic once $E(S_1) > 2 \cdot E(T_1)$, and endothermic for $E(S_1) < 2 \cdot E(T_1)$. Solid pentacene is an archetypal system exhibiting efficient exothermic singlet fission [27]. Interestingly, even in systems where it is endothermic, e.g. in tetracene [28–31], singlet fission can be very efficient and temperature-independent [30–33]. To explain such observations, coherent processes have been invoked [34, 35], a fact that we return to in Chap. 9. Singlet fission was first proposed in 1965 in order to explain the temperature dependence of the anthracene light-emission [36]. However, it is mostly in recent years that it has attracted a lot of attention, due to its promise

[1] There are exceptions to this rule, such as the organic radicals of Ref. [23] with a spin-doublet ground state, which allows these systems to achieve highly efficient light-emission.

for more efficient photovoltaic technologies. In a typical solar cell, the material absorbs light and an exciton is generated. The exciton then typically dissociates at an interface [37], forming a free electron and a hole that contribute towards electric current. Singlet exciton fission allows for the generation of two triplet excitons from a single absorbed photon, hence having the potential to produce up to double the amount of free charge carriers. However, the generated triplet excitons have low energies and cannot be used to increase the efficiency of solar cells based purely on organic semiconductors. Instead, one can use singlet fission for hybrid organic-inorganic devices. For example, consider silicon with a band gap of 1.1 eV. Photons of higher energy can still be absorbed, but their excess energy is lost in the form of heat. Instead one can couple silicon to a layer of the organic semiconductor tetracene. Tetracene absorbs at approximately 2.4 eV and undergoes efficient singlet fission. The generated triplets can then be transferred to silicon, since they have energies that lie very close to its electronic gap. In this way thermalisation losses are significantly reduced, and one can go beyond the theoretically predicted 'Shockley-Queisser' limit of 33% for the efficiency of single-junction solar cells, and up to 44% [38]. Indeed, efficient triplet transfer from tetracene to silicon was demonstrated in 2019 [39].

The triplet states that are produced from singlet fission are entangled into an overall spin-singlet triplet pair state, which is denoted as ^{1}TT. The spin-conserving character of this process allows it to be ultrafast, occurring in the femto- to picosecond timescale. The triplets that form the ^{1}TT state eventually lose coherence and become independent, however this process towards free triplets is much slower and can take nanoseconds in some cases [25].

The nature of the coupling between the S_1 and the ^{1}TT states has been a subject of debate in the literature. There are reports of both a direct transition between the two states, which requires a two-electron coupling, as well as a step-wise process involving charge transfer states as intermediates [25, 40–44]. These two mechanisms are schematically represented in Fig. 2.8, and it is the mediated mechanism which is considered to be dominant in most cases. This is due to the two-electron coupling between S_1 and ^{1}TT being at least an order of magnitude smaller than the one-electron couplings driving the mediated mechanism [40]. In the mediated fission scenario, the charge transfer states are not necessarily directly populated, but can facilitate the process as virtual intermediates in what is referred to as the super-exchange mechanism [40, 43].

Another factor that is increasingly recognised to be critical in the ultrafast timescale where singlet fission occurs, is the interaction of excitons with molecular and crystal vibrations [32, 35, 45–52] (key elements of the physics of vibrations in organic semiconductors are given in the following Sect. 2.3). Hence ultrafast processes, including singlet fission, provide an ideal platform to understand these effects, the microscopic mechanism of which has thus far remained elusive. In Chap. 7 we study the first step of the singlet fission process—ultrafast charge transfer—and find how vibrations can be used to increase its efficiency. Singlet fission is studied in Chaps. 8 and 9, where we reveal the microscopic mechanism and the role of exciton-vibration interactions in this process.

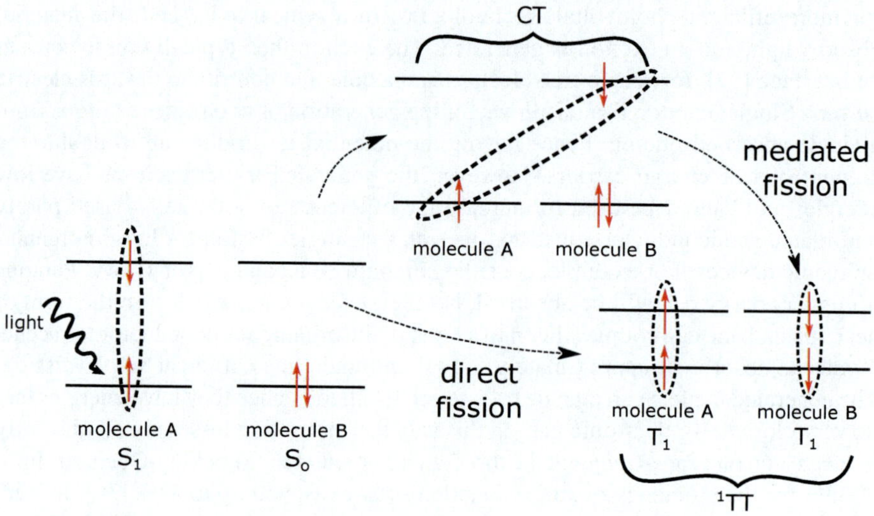

Fig. 2.8 Schematic representation of the singlet fission mechanism. A photon excites a singlet state S_1 on a molecule A, which transitions to a spin-singlet triplet-pair state ^1TT, with the two triplets localised on the neighbouring molecules A and B. The $S_1 \rightarrow {}^1TT$ transition proceeds either directly via a two-electron process, or via an intermediate charge transfer (CT) state

2.3 Vibrational Motion in Organic Semiconductors

We now consider the example of pentacene visualised in Fig. 2.9, in order to discuss the nature of vibrational motion in organic semiconductors. The conclusions that we draw in this section generalise to the vast majority of organic semiconductors.

There are roughly three categories into which vibrations of organic semiconductors can be classified. Firstly, there are vibrations that involve only pairs of atoms, e.g. two carbon atoms, or a carbon and a hydrogen. Then there are those vibrations which are collective in their nature, i.e. they involve larger groups of atoms, and up to every atom within a molecule. Finally, there are intermolecular vibrations, where two (or more) molecules move relative to each other.

The frequency of these various kinds of motions can be roughly modelled within a simple harmonic oscillator picture:

$$\omega = \sqrt{\frac{k}{\mu}}, \tag{2.14}$$

where k the spring constant and μ the effective mass of the oscillator. Collective motions of an entire molecule/pair of molecules have the lowest frequencies, due to the large mass of the oscillators involved. As we move towards more localised vibrations within a molecule, which only involve a small number of atoms, the

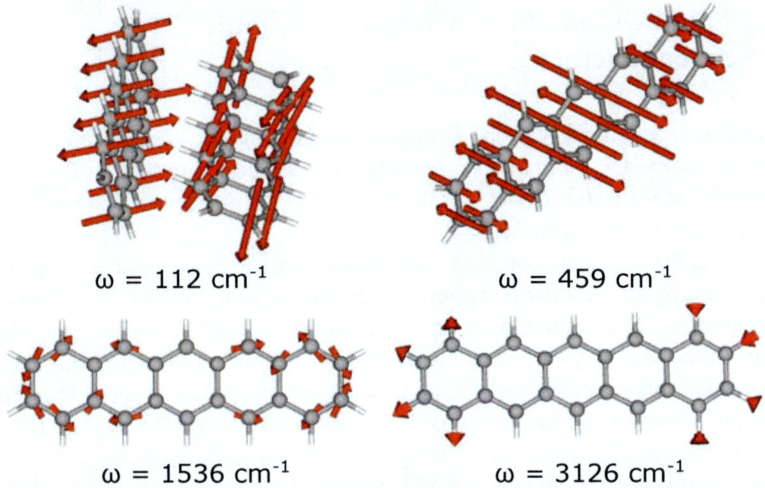

Fig. 2.9 A few characteristic vibrations of pentacene

frequency of the motion increases. When it comes to vibrations that only involve carbon atoms, the high frequency limit consists of motions that modulate the length of the chemical bond between a pair of carbon atoms. The spring constant k that appears in Eq. (2.14) encodes the strength of the chemical bond involved in the vibration. Hence a double bond carbon stretching motion has a higher frequency than the single bond stretching, and respectively the triple bond motion has a higher frequency than the double bond one. Finally, the relative motion of carbon and hydrogen is the highest frequency motion usually encountered in organic semiconductors.

At room temperature, the thermal energy of $25\,\text{meV} = 202\,\text{cm}^{-1}$ is sufficient to lead to a large displacement of the very low-frequency modes that arise due to the molecular nature of organic semiconductors. These slow and collective crystal vibrations lead to large variations of the distance and hence the electronic interactions between neighbouring molecules [53, 54]. Since excitons can spatially extend over several molecules, such low-energy phonons can have a strong influence on their properties, a point that we prove rigorously in Chap. 6. Organic semiconducting conjugated polymers also exhibit very low-frequency vibrations that are strongly coupled to their excitons [55]. However, we would like to emphasise that even high-frequency vibrational motions that are not thermally activated are important, due to zero-point nuclear quantum motion. These high-frequency oscillations have an energy of $E = \frac{\hbar\omega}{2}$, and can dominate the interaction with the exciton, e.g. upon absorption of a photon [56, 57], and during the non-equilibrium processes that follow. However, in Chap. 6 we show that even in the equilibrium state of several organic semiconductors, excitons can predominantly couple to vibrations that are only active due to quantum fluctuations.

2.4 Differences and Similarities to Inorganic Semiconductors

In the previous sections, various similarities and differences of organic semiconductors to traditional, inorganic semiconductors (which are those studied in classic textbooks of solid state physics [58]), become apparent. Here we summarise and further highlight some of these.

In metals, the dielectric constant is very large, with a 'sea' of electrons providing strong screening and leading to nearly free electronic motion. This is not the case in semiconductors. The relative permittivity of silicon, arguably the most widely used of semiconductors, is $\varepsilon_r = \frac{\varepsilon}{\varepsilon_o} = 11.4$, leading to a finite exciton binding energy of 15 meV [17]. This is still below the thermal energy of 25 meV at room temperature. As a result, it is most common to consider the motion of free electrons and free holes in these materials. The situation is different in organic semiconductors. Typical permittivity values range between 2 and 4 [59], leading to strong electron-hole attraction and exciton binding energies of several hundreds of meV. Hence it is generally the case that the absorption of a photon does not lead to free charges in these materials, but to bound exciton states.

The main characteristic of semiconducting materials, namely an energy gap between the highest occupied and lowest unoccupied states, is present in both organics and inorganics. In traditional semiconductors, the energy bands arise due to the periodicity of the atoms. Hence one speaks of **k**-dependent conduction and valence bands, where **k** a reciprocal lattice vector describing the state of a Bloch electron. This can also be the case in organic materials, in those cases that they exhibit a sufficient degree of periodicity. For example, the acene series of semiconductors is well-ordered in the solid state, and the picture of Bloch electrons described by a **k** vector holds. However, most organic molecules in the solid state form disordered films [60, 61]. In this case, there are no 'energy bands' in the sense of well-defined $E = E(\mathbf{k})$ relations, but only a distribution of orbital energies in real space. The HOMO-LUMO gap then takes the role of the energy gap.

Another important difference between organic and inorganic semiconductors is the nature of the vibrations the exhibit. As elaborated in Sect. 2.3, due to the molecular nature of organic semiconductors, their vibrations can be classified as intra- or intermolecular. Intermolecular vibrations are heavily activated at room temperature due to their very low frequencies, leading to a very strong interplay between electronic and vibrational degrees of freedom. This is not the case in inorganic atomic systems, where vibrational motion is often considered to act on the motion of electrons as a small perturbation [62]. It is partly this last property of organic semiconductors that makes it necessary to adopt approaches that go beyond perturbation theory for their description.

References

1. Pope M, Swenberg CE (1999) Electronic processes in organic crystals and polymers, 2nd edn. Oxford University Press, New York
2. Hund F (1925) Zur Deutung verwickelter Spektren, insbesondere der Elemente Scandium bis Nickel. Zeitschrift für Physik 34:345
3. Pauling L (1931) The nature of the chemical bond. Application of results obtained from the quantum mechanics and from a theory of paramagnetic susceptibility to the structure of molecules. J Am Chem Soc 53:1367
4. Hund F (1926) Zur Deutung einiger Erscheinungen in den Molekelspektren. Zeitschrift für Physik 36:657
5. Hund F (1927) Zur Deutung der Molekelspektren. I. Zeitschrift für Physik 40:742
6. Mulliken RS (1927) Electronic states and band spectrum structure in diatomic molecules IV. Hund's theory; second positive nitrogen and swan bands; alternating intensities. Phys Rev 29:637
7. Mulliken RS (1928) The assignment of quantum numbers for electrons in molecules. I. Phys Rev 32:186
8. Lennard-Jones LE (1929) The electronic structure of some diatomic molecules. Trans Faraday Soc 25:668
9. Hückel E. Quantentheoretische Beiträge zum Benzolproblem - I. Die Elektronenkonfiguration des Benzols und verwandter Verbindungen. Zeitschrift für Physik 70:204
10. Hoffmann R (1963) An extended Hückel theory. I. Hydrocarbons. J Chem Phys 39:1397
11. Rudolph P, Parr RG (1953) A semi-empirical theory of the electronic spectra and electronic structure of complex unsaturated molecules. I. J Chem Phys 21:466
12. Pople JA (1953) Electron interaction in unsaturated hydrocarbons. Trans Faraday Soc 49:1375
13. Chiang CK, Fincher CR, Park YW, Heeger AJ, Shirakawa H, Louis EJ, Gau SC, MacDiarmid AG (1977) Electrical conductivity in doped polyacetylene. Phys Rev Lett 39:1098
14. Fincher CR, Peebles DL, Heeger AJ, Druy MA, Matsumura Y, MacDiarmid AG, Shirakawa H, Ikeda S (1978) Anisotropic optical properties of pure and doped polyacetylene. Solid State Commun 27:489
15. Tiago ML, Michael R, Louie SG (2004) Bound excitons and optical properties of bulk trans-polyacetylene. Phys Rev B 70:193204
16. Schuster R, Knupfer M, Berger H (2007) Exciton band structure of pentacene molecular solids: breakdown of the Frenkel exciton model. Phys Rev Lett 98:037402
17. Green Martin A (2013) Improved value for the silicon free exciton binding energy. AIP Adv 3.112104
18. Frenkel J (1931) On the transformation of light into heat in solids. I. Phys Rev 37:17
19. Wannier Gregory H (1937) The structure of electronic excitation levels in insulating crystals. Phys Rev 52:191
20. Cohen ML, Louie SG (2016) Fundamentals of condensed matter physics. Cambridge University Press, Cambridge
21. Köhler A, Bässler H (2009) Triplet states in organic semiconductors. Mater Sci Eng R Rep 66:71
22. Dirac PAM, Bohr NHD (1927) The quantum theory of the emission and absorption of radiation. Proc R Soc A 114:243
23. Ai X, Evans EW, Dong S, Gillett AJ, Guo H, Chen Y, Hele TJH, Friend RH, Li F (2018) Efficient radical-based light-emitting diodes with doublet emission. Nature 563:536
24. Marian CM (2012) Spin-orbit coupling and intersystem crossing in molecules. Wiley Interdisc Rev Comput Mole Sci 2:187
25. Smith MB, Michl J (2010) Singlet fission. Chem Rev 110:6891
26. Rao A, Friend RH (2017) Harnessing singlet exciton fission to break the Shockley-Queisser limit. Nat Rev Mater 2:17063
27. Rao A, Wilson MWB, Albert-Seifried S, Di Pietro R, Friend RH (2011) Photophysics of pentacene thin films: the role of exciton fission and heating effects. Phys Rev B 84:195411

28. Swenberg CE, Stacy WT (1968) Bimolecular radiationless transitions in crystalline tetracene. Chem Phys Lett 2:327
29. Burdett JJ, Bardeen CJ (2012) Quantum beats in crystalline tetracene delayed fluorescence due to triplet pair coherences produced by direct singlet fission. J Am Chem Soc 134:8597
30. Wilson MW, Rao A, Johnson K, Gélinas S, Di Pietro R, Clark J, Friend RH (2013) Temperature-independent singlet exciton fission in tetracene. J Am Chem Soc 135:16680
31. Tayebjee MJY, Clady RGCR, Schmidt TW (2013) The exciton dynamics in tetracene thin films. Phys Chem Chem Phys 15:14797
32. Stern HL, Cheminal A, Yost SR, Broch K, Bayliss SL, Chen K, Tabachnyk M, Thorley K, Greenham N, Hodgkiss JM, Anthony J, Head-Gordon M, Musser AJ, Rao A, Friend RH (2017) Vibronically coherent ultrafast triplet-pair formation and subsequent thermally activated dissociation control efficient endothermic singlet fission. Nat Chem 9:1205
33. Burdett JJ, Gosztola D, Bardeen CJ (2011) The dependence of singlet exciton relaxation on excitation density and temperature in polycrystalline tetracene thin films: kinetic evidence for a dark intermediate state and implications for singlet fission. J Chem Phys 135:214508
34. Chan WL, Ligges M, Zhu XY (2012) The energy barrier in singlet fission can be overcome through coherent coupling and entropic gain. Nat Chem 4:840
35. Kiyoshi M, Yuki K, Kazuya W, Toshiki S, Shota T, Shunsuke T, Jun T, Takeshi Y, Yoshiyasu M (2017) Coherent singlet fission activated by symmetry breaking. Nat Chem 9:983
36. Singh S, Jones WJ, Siebrand W, Stoicheff BP, Schneider WG (1965) Laser generation of excitons and fluorescence in anthracene crystals. J Chem Phys 42:330
37. Clarke TM, Durrant JR (2010) Charge photogeneration in organic solar cells. Chem Rev 110:6736
38. Hanna MC, Nozik AJ (2006) Solar conversion efficiency of photovoltaic and photoelectrolysis cells with carrier multiplication absorbers. J Appl Phys 100:074510
39. Einzinger M, Tony W, Kompalla JF, Smith HL, Perkinson CF, Nienhaus L, Wieghold S, Congreve DN, Kahn A, Bawendi MG, Baldo MA (2019) Sensitization of silicon by singlet exciton fission in tetracene. Nature 571:90
40. Berkelbach TC, Hybertsen MS, Reichman DR (2013) Microscopic theory of singlet exciton fission. II. Application to pentacene dimers and the role of superexchange. J Chem Phys 138:114103
41. Zimmerman PM, Bell F, Casanova D, Head-Gordon M (2011) Mechanism for singlet fission in pentacene and tetracene: From single exciton to two triplets. J Am Chem Soc 133:19944
42. Smith MB, Michl J (2013) Recent advances in singlet fission. Ann Rev Phys Chem 64:361
43. Berkelbach TC, Hybertsen MS, Reichman DR (2013) Microscopic theory of singlet exciton fission. I. General formulation. J Chem Phys 138:114102
44. Zimmerman PM, Zhang Z, Musgrave CB (2010) Singlet fission in pentacene through multi-exciton quantum states. Nat Chem 2:648
45. Musser AJ, Liebel M, Schnedermann C, Wende T, Kehoe TB, Rao A, Kukura P (2015) Evidence for conical intersection dynamics mediating ultrafast singlet exciton fission. Nat Phys 11:352
46. Bakulin AA, Morgan SE, Kehoe TM, Wilson MWB, Chin AW, Zigmantas D, Egorova D, Rao A (2016) Real-time observation of multiexcitonic states in ultrafast singlet fission using coherent 2D electronic spectroscopy. Nat Chem 8:16
47. Tempelaar R, Reichman DR (2018) Vibronic exciton theory of singlet fission. III. How vibronic coupling and thermodynamics promote rapid triplet generation in pentacene crystals. J Chem Phys 148:244701
48. Fujihashi Y, Chen L, Ishizaki A, Wang J, Zhao Y (2017) Effect of high-frequency modes on singlet fission dynamics. J Chem Phys 146:044101
49. Tamura H, Huix-Rotllant M, Burghardt I, Olivier Y, Beljonne D (2015) First-principles quantum dynamics of singlet fission: coherent versus thermally activated mechanisms governed by molecular π stacking. Phys Rev Lett 115:107401
50. Feng X, Krylov AI (2016) On couplings and excimers: lessons from studies of singlet fission in covalently linked tetracene dimers. Phys Chem Chem Phys 18:7751

51. Huang Z, Fujihashi Y, Zhao Y (2017) Effect of off-diagonal exciton-phonon coupling on intramolecular singlet fission. J Phys Chem Lett 8:3306
52. Schröder FAYN, Turban DHP, Musser AJ, Hine NDM, Chin AW (2019) Tensor network simulation of multi-environmental open quantum dynamics via machine learning and entanglement renormalisation. Nat Commun 10:1062
53. Troisi A, Orlandi G, Anthony JE (2005) Electronic interactions and thermal disorder in molecular crystals containing cofacial pentacene units. Chem Mater 17:5024
54. Troisi A, Orlandi G (2006) Dynamics of the intermolecular transfer integral in crystalline organic semiconductors. J Phys Chem A 110:4065
55. Mahrt RF, Bässler H (1991) Light and heavy excitonic polarons in conjugated polymers. Synthetic Metals 45:107
56. Spano FC (2010) The spectral signatures of frenkel polarons in H- And J-aggregates. Accounts Chem Res 43:429
57. Yamagata H, Spano FC (2011) Vibronic coupling in quantum wires: applications to polydiacetylene. J Chem Phys 135:054906
58. Ashcroft NW, David Mermin N (1976) Solid state physics. Holt, Rinehart & Winston, New York, London
59. Torabi S, Jahani F, Van Severen I, Kanimozhi C, Patil S, Havenith RWA, Chiechi RC, Lutsen L, Vanderzande DJM, Cleij TJ, Hummelen JC, Jan Anton Koster L (2015) Strategy for enhancing the dielectric constant of organic semiconductors without sacrificing charge carrier mobility and solubility. Adv Functional Mater 25:150
60. Neumann T, Danilov D, Lennartz C, Wenzel W (2013) Modeling disordered morphologies in organic semiconductors. J Comput Chem 34:2716
61. Forrest SR (2004) The path to ubiquitous and low-cost organic electronic appliances on plastic. Nature 428:911
62. Saha K, Garate I (2014) Phonon-induced topological insulation. Phys Rev B 89:205103

Chapter 3
The Time-Dependent Quantum Mechanical Problem

This chapter aims to introduce a number of concepts which are important for modelling the quantum dynamics of an exciton system interacting with an environment of molecular and/or crystal vibrations. We start in Sect. 3.1 from discussing ways of formulating the problem of obtaining the time-dependent properties of a system, based on the so-called density matrix formalism. We then proceed in Sect. 3.2 to examine the case of a system of interest interacting with another, potentially large, system, which we shall call 'the environment'. In Sect. 3.3 we present a number of approximations which are employed to derive quantum master equations that are commonly used to compute the time evolution of large systems. We briefly discuss these approximations in Sect. 3.4, highlighting the fact that they often break down in organic semiconductors. It is therefore necessary to develop and use a methodology that can overcome these shortcomings. This methodology is based on tensor networks, and we discuss it in detail in Chaps. 7 and 8. We refer the reader interested in more detail on the material of this chapter to Ref. [1].

3.1 Density Matrix Formalism

Let us consider a statistical mixture of quantum states $|\Psi_n\rangle$, each of which has a statistical weight W_n, with n a positive integer. The density operator describing such a mixture is defined as:

$$\rho = \sum_n W_n |\Psi_n\rangle \langle\Psi_n| . \tag{3.1}$$

Each of the states $|\Psi_n\rangle$ in the above sum is a pure state, in other words it can always be written as a superposition of a set of basis states $|\Phi_m\rangle$ of our choice:

© The Author(s), under exclusive license to Springer Nature Switzerland AG 2021

A. M. Alvertis, *On Exciton–Vibration and Exciton–Photon Interactions in Organic Semiconductors*, Springer Theses,
https://doi.org/10.1007/978-3-030-85454-6_3

$$|\Psi_n\rangle = \sum_m \alpha_m^{(n)} |\Phi_m\rangle . \tag{3.2}$$

In this basis, the density operator is rewritten as:

$$\rho = \sum_{mm'n} W_n \alpha_m^{(n)} \alpha_{m'}^{(n)*} |\Phi_m\rangle \langle \Phi_{m'}| , \tag{3.3}$$

and its matrix representation in the basis of $|\Phi_m\rangle$ is:

$$\langle \Phi_i | \rho | \Phi_j \rangle = \sum_n W_n \alpha_i^{(n)} \alpha_j^{(n)*} . \tag{3.4}$$

Equation (3.4) defines the *density matrix* within a specific basis set.

There are a number of properties of the density matrix that constitute it a very useful quantity to calculate when considering the dynamics of a quantum system, so let us elaborate on some of these:

- The density matrix is Hermitian, since the matrix elements of Eq. (3.4) satisfy the condition:
$$\langle \Phi_i | \rho | \Phi_j \rangle = \langle \Phi_j | \rho | \Phi_i \rangle^* \tag{3.5}$$
- The diagonal elements ρ_{mm} of the density matrix give the probability of finding the system in the state $|\Phi_m\rangle$:
$$\rho_{mm} = \sum_n W_n |\alpha_m^{(n)}|^2 \tag{3.6}$$
This becomes evident from the fact that W_n is the probability of finding the system in state $|\Psi_n\rangle$, which in turn has probability $|\alpha_m^{(n)}|^2$ to be found in the state $|\Phi_m\rangle$. As a consequence, the sum of these diagonal elements is equal to one: $\mathrm{tr}(\rho) = 1$.
- The off-diagonal elements ρ_{mn} of the density matrix indicate whether the system is in a coherent superposition or not. The matrix element ρ_{mn} is often called the *coherence* between states $|\Phi_m\rangle$ and $|\Phi_n\rangle$. For a system in the pure state $|\Psi\rangle = \alpha_1 |\Phi_1\rangle + \alpha_2 |\Phi_2\rangle$, $\rho_{12} = \alpha_1 \alpha_2^*$. The coherence is a measure of the quantum versus classical character of a system. In the superposition state $|\Psi\rangle = \frac{1}{\sqrt{2}}(|\Phi_1\rangle + |\Phi_2\rangle)$ of such a two-level system, the coherence obtains its maximal value $\rho_{12} = 0.5$, whereas $\rho_{12} = 0$ in the case where the system has collapsed into one of the two basis states. Even if the system is in a mixture of different quantum states, which however are not coherent superpositions of basis states, the coherence will still be equal to zero, as it becomes obvious from Eq. (3.4). This is called an *incoherent mixture*. The coherence of a system depends on the specific choice of basis.
- The relation:
$$\mathrm{tr}(\rho^2) \leq (\mathrm{tr}(\rho))^2 \tag{3.7}$$

holds, with the equality being a sufficient and necessary condition that the system described by ρ is in a pure state.

- The expectation value of any operator A is equal to the trace of ρ times A, since:

$$\begin{aligned}\langle A\rangle &= \sum_{mm'}\sum_{n} W_n \alpha_m^{(n)} \alpha_{m'}^{(n)*} \langle\Phi_{m'}|\, A\, |\Phi_m\rangle \\ &= \sum_{mm'} \langle\Phi_m|\, \rho\, |\Phi_{m'}\rangle \langle\Phi_{m'}|\, A\, |\Phi_m\rangle = \mathrm{tr}(\rho A)\end{aligned} \tag{3.8}$$

At this stage, it is important to note that since in quantum mechanics a system is fully described by the expectation values of a set of mutually commuting operators, and these expectation values may be obtained through relation (3.8), consequently the density matrix contains all physically significant information of the system.

Now let us see how the density matrix may be used in the context of considering the time evolution of a quantum system. For a general time-dependent Hamiltonian $\mathcal{H}(t)$, the evolution of the system is governed by the Schrödinger equation:

$$i\frac{\partial\,|\Psi(t)\rangle}{\partial t} = \mathcal{H}(t)\,|\Psi(t)\rangle\,, \tag{3.9}$$

where we remind that throughout this thesis we work in units of $\hbar = 1$. For a time-independent Hamiltonian this admits the simple solution:

$$|\Psi(t)\rangle = \exp(-i\mathcal{H}t)\,|\Psi(0)\rangle\,, \tag{3.10}$$

however once we have an explicit time dependence in $\mathcal{H}$, this is not the case any more. Generally, $|\Psi(t)\rangle$ may be obtained by operating on $|\Psi(0)\rangle$ with a time evolution operator $U(t)$:

$$|\Psi(t)\rangle = U(t)\,|\Psi(0)\rangle\,. \tag{3.11}$$

By substituting Eqs. (3.11) into (3.9), we obtain:

$$i\frac{\partial U(t)}{\partial t}\,|\Psi(0)\rangle = \mathcal{H}(t)U(t)\,|\Psi(0)\rangle\,, \tag{3.12}$$

which holds for any $|\Psi(0)\rangle$, and therefore:

$$i\frac{\partial U(t)}{\partial t} = \mathcal{H}(t)U(t). \tag{3.13}$$

Similarly, for the Hermitian conjugate:

$$-i\frac{\partial U^{\dagger}(t)}{\partial t} = U^{\dagger}(t)\mathcal{H}(t). \tag{3.14}$$

Suppose now that at time $t = 0$ a system in a mixed state is represented by the density operator:

$$\rho(0) = \sum_n W_n \left|\Psi(0)_n\right\rangle \left\langle\Psi(0)_n\right| . \tag{3.15}$$

According to the time evolution of Eq. (3.11), at times t the operator becomes:

$$\begin{aligned}\rho(t) &= \sum_n W_n \left|\Psi(t)_n\right\rangle \left\langle\Psi(t)_n\right| \\ &= \sum_n W_n U(t) \left|\Psi(0)_n\right\rangle \left\langle\Psi(0)_n\right| U^\dagger(t),\end{aligned} \tag{3.16}$$

which may be rewritten as:

$$\rho(t) = U(t)\rho(0)U^\dagger(t). \tag{3.17}$$

We can rewrite Eq. (3.17) by differentiating it with respect to time and making use of Eqs. (3.13) and (3.14):

$$\begin{aligned}i\frac{\partial\rho(t)}{\partial t} &= i\frac{\partial U(t)}{\partial t}\rho(0)U^\dagger(t) + iU(t)\rho(0)\frac{\partial U^\dagger(t)}{\partial t} \\ &= \mathcal{H}(t)U(t)\rho(0)U^\dagger(t) - U(t)\rho(0)U^\dagger(t)\mathcal{H}(t).\end{aligned} \tag{3.18}$$

Finally, by using Eq. (3.17), we obtain:

$$i\frac{\partial\rho(t)}{\partial t} = [\mathcal{H}(t), \rho(t)] . \tag{3.19}$$

Equation (3.19) for the time dependence of the density matrix is commonly referred to as the Liouville equation. This may also be rewritten as an integral equation:

$$\rho(t) = \rho(0) - i\int_0^t \left[\mathcal{H}(t'), \rho(t')\right] dt' \tag{3.20}$$

All the preceding discussion has been based on the *Schrödinger picture* of quantum mechanics, within which the wavefunction contains all the time dependence of a system, according to Eq. (3.11). An alternative representation of the time-dependent problem, which we shall use later on in this chapter, is given by the so-called *interaction picture*. Consider a system described by the Hamiltonian $\mathcal{H}_o$, perturbed by a time-dependent external potential $V(t)$:

$$\mathcal{H} = \mathcal{H}_o + V(t). \tag{3.21}$$

Within this picture, state vectors obey the equation:

$$i\frac{\partial \left|\Psi(t)_I\right\rangle}{\partial t} = V(t)_I \left|\Psi(t)_I\right\rangle, \tag{3.22}$$

where the interaction picture state vectors $|\Psi(t)_I\rangle$ are obtained from their Schrödinger picture counterparts by removing the time-dependence created by $\mathcal{H}_o$:

$$\left|\Psi(t)_I\right\rangle = \exp(i\mathcal{H}_o t)\left|\Psi(t)\right\rangle, \tag{3.23}$$

and for the potential term:

$$V(t)_I = \exp(i\mathcal{H}_o t)V(t)\exp(-i\mathcal{H}_o t). \tag{3.24}$$

Similarly, all operators $A(t)$ in the Schrödinger picture are transformed in the interaction picture:

$$A(t)_I = \exp(i\mathcal{H}_o t)A(t)\exp(-i\mathcal{H}_o t), \tag{3.25}$$

The time-dependence of the state vectors in the interaction picture is created entirely by the external potential $V(t)$, as it becomes evident from Eq. (3.22). The interaction picture can be convenient in cases where $V(t)$ is a small perturbation of the system. In these cases, $|\Psi(t)_I\rangle$ varies slowly in time, meaning that (3.22) can be solved approximately using time-dependent perturbation theory, which makes it more practical than Eq. (3.9).

3.2 The System and Its Environment

The Liouville Eqs. (3.19) or (3.20) determines the time evolution of a generic quantum system described by a density matrix $\rho(t)$. However, the situation with which one is often faced, is that of being interested in the dynamics of a quantum system, which is interacting with a second, potentially very large quantum system. We will refer to the quantum system of interest simply as the 'system' (S), and the large system with which it interacts as its 'environment' (E). This situation is of particular importance for this thesis, since we are often concerned with an exciton system resulting from the absorption of light, which is interacting with a very large number of molecular or crystal vibrations, i.e. a large vibrational environment. We are in most cases interested in the observables of the exciton system, while the environment remains unobserved.

Let us denote the basis states of the system as $\left|\varphi_j\right\rangle$, and those of the environment as $|\Phi_i\rangle$. The collective state of the system and the environment will then be denoted as $\left|\Phi_i\varphi_j\right\rangle$. Consider now an operator A corresponding to an observable of the system, which only acts on the states $\left|\varphi_j\right\rangle$. The matrix representation of this operator is:

$$\left\langle \Phi_{i'}\varphi_{j'} \right| A \left| \Phi_i \varphi_j \right\rangle = \left\langle \varphi_{j'} \right| A \left| \varphi_j \right\rangle \delta_{i'i}, \tag{3.26}$$

and by using this and Eq. (3.8), we obtain the expectation value of A at times t:

$$\begin{aligned} \langle A(t) \rangle = \mathrm{tr}(\rho(t)A) &= \sum_{i'ij'j} \left\langle \Phi_{i'}\varphi_{j'} \right| \rho(t) \left| \Phi_i \varphi_j \right\rangle \left\langle \Phi_i \varphi_j \right| A \left| \Phi_{i'}\varphi_{j'} \right\rangle \\ &= \sum_{j'j} [\sum_i \left\langle \Phi_i \varphi_{j'} \right| \rho(t) \left| \Phi_i \varphi_j \right\rangle] \left\langle \varphi_j \right| A \left| \varphi_{j'} \right\rangle \end{aligned} \tag{3.27}$$

We define:

$$\left\langle \varphi_{j'} \right| \rho_S(t) \left| \varphi_j \right\rangle = \sum_i \left\langle \Phi_i \varphi_{j'} \right| \rho(t) \left| \Phi_i \varphi_j \right\rangle, \tag{3.28}$$

the *reduced density matrix* of the system. This is obtained by tracing over the variables of the environment, essentially constituting a projection of the density matrix onto the subsystem of interest. Based on this definition, Eq. (3.27) can be rewritten:

$$\begin{aligned} \langle A(t) \rangle &= \sum_{j'j} \left\langle \varphi_{j'} \right| \rho_S(t) \left| \varphi_j \right\rangle \left\langle \varphi_j \right| A \left| \varphi_{j'} \right\rangle \\ &= \mathrm{tr}(\rho_S(t)A). \end{aligned} \tag{3.29}$$

Similar to Eqs. (3.8, 3.29) means that the reduced density matrix $\rho_S(t)$ contains all information on the system S.

Now let us consider the Hamiltonian operator of the enlarged system, consisting of our system of interest S and its environment E:

$$\mathcal{H} = \mathcal{H}_o(S) + \mathcal{H}_o(E) + V, \tag{3.30}$$

where the terms $\mathcal{H}_o$ refer to the free motion of the two systems, and V describes their interaction. Because of this finite interaction, there is no Hamiltonian operator which can describe the dynamics of the system S alone.

One can still describe the time evolution of the system S by tracing over the variables of the environment E (denoted as tr_E) and forming the reduced density matrix in Eq. (3.17):

$$\rho_S(t) = \mathrm{tr}_E(U(t)\rho(0)U^\dagger(t)), \tag{3.31}$$

or equivalently:

$$\frac{\partial \rho_S(t)}{\partial t} = -i\,\mathrm{tr}_E([\mathcal{H}, \rho(t))]). \tag{3.32}$$

It is interesting to note that in Eq. (3.31), the operation of taking the trace over the variables of the environment E, prevents one from obtaining $\rho(0)$ from $\rho(t)$ through the inverse transformation $U^{\dagger}(t)\rho(t)U(t)$. This is a hallmark of irreversible behaviour, which results from the interaction with the unobserved system E.

Such a system S having a non-negligible mutual interaction with an unobserved environment E is called an *open quantum system*. As we discussed above, the time evolution of such a system cannot be described by the Liouville Eqs. (3.17) or (3.19), but by Eqs. (3.31) and (3.32) that lead to irreversible behaviour. Additionally, according to the principle of non-separability [2], the interaction results in a mixed state for the open system, even if initially it was found in a pure state. This can be understood as an increase of the so-called von Neumann entropy:

$$S_{\mathrm{vNE}} = -\mathrm{tr}(\rho \ln \rho) = -\sum_{i} p_i \ln p_i, \tag{3.33}$$

where p_i the probability of finding the system in the state i, as obtained by Eq. (3.6). This increase in entropy underlines the irreversible character of the open quantum system dynamics.

Of course, it is always possible to describe the enlarged system, consisting of the combination of S and E, as a single *closed* quantum mechanical system. This isolated system is described by a Hamiltonian $\mathcal{H}$ and a time evolution operator $U(t)$, while the Liouville Eq. (3.19) holds. Naturally, for systems in contact with a large environment this approach requires one to perform the time evolution of an enlarged system of what has often been thought of as forbidding size, since the size of the associated Hilbert space scales exponentially with the number of basis states. Therefore, in most cases the avenue of open quantum systems is deemed more appropriate. We will show however in Chap. 7 that by using tensor network methods, one can describe the combined dynamics of an exciton-vibration system of large organic molecules, at a low computational cost. This formalism has the additional advantage of allowing us to go beyond a number of very common approximations which are used within the open quantum system framework, and which are described in the following section.

3.3 Approximate Quantum Master Equations

Let us now consider a number of approximations which are commonly used to arrive at different quantum master equations that govern the time evolution of a system S interacting with an unobserved environment E. The time evolution of the density matrix $\rho(t)$ of the combined system, described by the Hamiltonian (3.30), is given by (3.19), or equivalently by (3.20). We insert (3.20) into (3.19) and obtain:

$$\dot{\rho}(t) = -i\left[\mathcal{H}(t), \rho(0)\right] + \int_0^t dt' \left[\mathcal{H}(t), \left[\mathcal{H}(t'), \rho(t')\right]\right]. \tag{3.34}$$

We now switch to the interaction picture, within which the time evolution of the density matrix is governed by the interaction potential V. Since we are interested in the observables of the system S, we trace over the the variables of E and obtain the reduced density matrix in the interaction picture:

$$\dot{\rho}(t)_{SI} = -i\mathrm{tr}_E([V(t)_I, \rho(0)_I]) + \int_0^t dt'\mathrm{tr}_E([V(t)_I, [V(t')_I, \rho(t')_I]]), \quad (3.35)$$

where generally an operator A in the interaction picture is related to the Schrödinger picture according to Eq. (3.25).

If we assume the interaction V between S and E to be switched on at $t = 0$, then prior to this the two systems are uncorrelated (the associated Hilbert spaces disjoint) and the total density matrix is given by the direct product:

$$\rho(0) = \rho(0)_S \otimes \rho(0)_E. \quad (3.36)$$

We now proceed to make the first key approximation towards the derivation of a quantum master equation:

- **Approximation I**: The Born approximation
  We assume the coupling of the system S to the environment E to be weak, so that the influence of the interaction on E is small. Therefore E can always be represented by $\rho(0)_E$ and the total density matrix is written as:

$$\rho(t)_I = \rho(t)_{SI}\rho(0)_E. \quad (3.37)$$

  This is equivalent to assuming that the environment is very large, therefore the effects of its interaction with the system dissipate away quickly. This approximation is intimately connected to the Born-Oppenheimer approximation discussed in Chap. 4.

Substituting the approximate density matrix (3.37) in Eq. (3.35), we obtain:

$$\dot{\rho}(t)_{SI} = -i\mathrm{tr}_E([V(t)_I, \rho(0)_{S\rho}(0)_E]) + \int_0^t dt'\mathrm{tr}_E([V(t)_I, [V(t')_I, \rho(t')_{SI}\rho(0)_E]]), \quad (3.38)$$

Due to the appearance of $\rho(t')_{SI}$ in the integral of Eq. (3.38), the value of $\rho(t)$ depends on the past behaviour of the system, from $t' = 0$ to $t' = t$. This is of particular relevance to the second common approximation:

- **Approximation II**: The Markov approximation
  We assume that the environment E loses memory of its past interactions with the system S within a short timescale τ. As a result, only terms between the times $t' = t - \tau$ and $t' = t$ contribute to the integral of Eq. (3.38), and terms outside this interval have a negligible contribution. We can therefore extend the integration to infinity without introducing a large error. Additionally, if $1/\gamma$ a characteristic timescale for the system to show appreciable change at a macroscopic level (e.g. a decay or damping time), then in the case that $\tau << 1/\gamma$, we can approximate:

$$\rho(t')_{SI} = \rho(t)_{SI}, \tag{3.39}$$

  in the integrand of Eq. (3.38), as long as we are not interested in detailed microscopic behaviour.

By using the Markov approximation, Eq. (3.38) gives us the first *Markovian master equation* of this section:

$$\begin{aligned} \dot{\rho}(t)_{SI} &= -i\mathrm{tr}_E([V(t)_I, \rho(0)_S\rho(0)_E]) \\ &\quad + \int_0^\infty dt'\mathrm{tr}_E([V(t)_I, [V(t')_I, \rho(t)_{SI}\rho(0)_E]]). \end{aligned} \tag{3.40}$$

This time-local master equation is much easier to integrate compared to (3.38).

From the Markovian master Eq. (3.40), it is possible to derive different master equations, which in some cases require additional approximations. Here we will discuss two such master equations, the **Bloch-Redfield** and **Lindblad** master equations, which have been used in the context of modelling ultrafast photoinduced dynamics in organic semiconductors [3–5]. We will not provide the full derivation of these master equations, but only the most essential steps. For more detail, we refer the interested reader to references [6, 7].

First we assume that the interaction V between the system S and the environment E can be written as:

$$V = \sum_j A_j \otimes B_j, \tag{3.41}$$

where the operators A_j and B_j only act on the variables of S and E respectively, and commute with each other. The system operators may be decomposed:

$$A_j = \sum_\omega A_j(\omega), \tag{3.42}$$

with each of the operators $A_j(\omega)$ mediating a transition from state n to state m: $\omega = \varepsilon_n - \varepsilon_m$. By using the above, Eq. (3.40) may be rewritten as:

$$\dot{\rho}(t)_{SI} = \sum_{\omega,\omega'} \sum_{j} e^{i(\omega-\omega')t} \Gamma_j(\omega)$$
$$[A_j(\omega)\rho(t)_{SI} A_j(\omega')^\dagger - A_j(\omega')^\dagger A_j(\omega)\rho(t)_{SI}] + h.c., \tag{3.43}$$

with:

$$\Gamma_j(\omega) = \int_0^\infty dt' e^{i\omega t'} \mathrm{tr}_E(B_j(t')^\dagger B_j(0)\rho(0)_E)$$
$$= \frac{1}{2}\gamma_j(\omega) + i\lambda_j(\omega). \tag{3.44}$$

Equation (3.43) is the *Bloch-Redfield* master equation in the interaction picture. Here $\gamma_j(\omega) = \int_{-\infty}^{\infty} dt' e^{i\omega t'} \mathrm{tr}_E(B_j(t')^\dagger B_j(0)\rho(0)_E)$ assumes real positive values and induces transitions between states of frequencies ω and ω' appearing in Eq. (3.43). This is analogous to the case of a perturbation inducing a transition between two states, as described by Fermi's golden rule Eq. (2.11). The imaginary part $\lambda_j(\omega)$ results in a shift of the system eigenfrequencies through the so-called Lamb shift Hamiltonian:

$$\mathcal{H}_{LS} = \sum_{\omega,j} \lambda_j(\omega) A_j(\omega)^\dagger A_j(\omega). \tag{3.45}$$

One of the problems associated with the Bloch-Redfield master equation is that if the system-environment coupling becomes strong, then it can give rise to negative values for the reduced density matrix. One may overcome this issue by employing an additional approximation:

- **Approximation III**: The Secular approximation
  Within this approximation, we neglect any terms in Eq. (3.43) representing non-resonant processes. Hence terms containing $e^{i(\omega-\omega')t}$ for $\omega \neq \omega'$ are ignored, due to exhibiting fast oscillations compared to the typical relaxation time of the system.

By using the secular approximation, we obtain the *Lindblad* master Eq. [7]:

$$\dot{\rho}(t)_{SI} = -i[\mathcal{H}_{LS}, \rho(t)_{SI}] + \sum_{\omega,j} \gamma_j(\omega)$$
$$[A_j(\omega)\rho(t)_{SI} A_j(\omega)^\dagger - \frac{1}{2}\{A_j(\omega)^\dagger A_j(\omega), \rho(t)_{SI}\}] + h.c., \tag{3.46}$$

3.4 Discussion of the Approximations

When deriving the Markovian master equations of the previous section, we made use of three different approximations: the Born approximation (weak system-environment coupling), the Markov approximation (negligible memory effects on the system) and the secular approximation (contributions only from resonant terms). We will however see in following chapters that all of these approximations can often break down when studying the time-dependent response of organic systems to a photon excitation. Very strong exciton-vibration interactions are found for the systems studied in all of the Chaps. 5–9, making the validity of the Born approximation questionable. Furthermore, the closely-related Born-Oppenheimer approximation which is discussed in detail in the following chapter, breaks down in the vicinity of a crossing between potential energy surfaces, which is present in the process of singlet fission in the two different systems studied in Chaps. 8 and 9. Memory effects are found to be critical in the case of using excess energy to enhance charge transfer in Chap. 7, while vibrations that are not in resonance with electronic transitions can lead to significant effects. We thus conclude that the Markov and Secular approximations also fail to capture the essential physics of these systems.

With the above we would like to highlight that the commonly employed quantum master equations can often not be applied in the context of light-induced processes in organic systems, making it necessary to use a more advanced methodology. For the solution of this time-dependent problem, tensor networks are our method of choice. We present these methods within the context of ultrafast processes in Chap. 7, building upon the developments presented in [6], extending them in order to describe the effects of excess energy excitation and study charge transfer. We then apply these methods to reveal the full quantum mechanical dynamics of a large molecule during the process of singlet fission in Chap. 8.

References

1. Blum K (2012) Density matrix theory and applications/Karl Blum, 3rd edn. In: Springer series on atomic, optical, and plasma physics, 64. Springer, Heidelberg, London
2. Espagnat B (1971) Conceptual foundations of quantum mechanics. Mathematical physics monograph series. W. A. Benjamin, Menlo Park, Calif
3. Chan WL, Berkelbach TC, Provorse MR, Monahan NR, Tritsch JR, Hybertsen MS, Reichman DR, Gao J, Zhu XY (2013) The quantum coherent mechanism for singlet fission: Experiment and theory. Accounts Chem Res 46:1321
4. Xie X, Santana-Bonilla A, Fang W, Liu C, Troisi A, Ma H (2019) Exciton-Phonon interaction model for singlet fission in prototypical molecular crystals. J Chem Theor Comput 15:3721
5. Berkelbach TC, Hybertsen MS, Reichman DR (2013) Microscopic theory of singlet exciton fission. II. Application to pentacene dimers and the role of superexchange. J Chem Phys 138:114103
6. Schröder FAYN (2017) Tensor network states simulations of exciton-phonon quantum dynamics for applications in artificial light-harvesting. PhD thesis, University of Cambridge
7. Manzano Daniel (2020) A short introduction to the Lindblad master equation. AIP Advances 10:025106

Chapter 4
Modelling of the Electronic and Vibrational Structure

Check for updates

4.1 The Many-Body Hamiltonian and the Born-Oppenheimer Approximation

For systems with a time-independent Hamiltonian, the Schrödinger equation dictates a simple time dependence that manifests itself as phase factors multiplying the solutions of the time-independent Schrödinger equation:

$$\mathcal{H}\,|\Psi\rangle = E\,|\Psi\rangle\,, \tag{4.1}$$

Obtaining the solutions of the time-independent Schrödinger equation is a prerequisite for understanding time-dependent properties. In this chapter we will discuss a number of methods that are used throughout this thesis for solving Eq. (4.1) in material systems and organic semiconductors in particular.

We start by considering the most general form for the above Hamiltonian when describing a material system. Since this thesis is concerned with organic molecules consisting predominantly of hydrogen and carbon, we ignore relativistic corrections to the Hamiltonian. The magnitude of these corrections is proportional to Z^4 [1], where Z the atomic number, and therefore become important once a system contains heavy elements. As a result, the Hamiltonian may be written by considering the classical kinetic and potential terms for electrons i and nuclei α:

$$\begin{aligned}\mathcal{H} = &\frac{1}{2}\sum_i \mathbf{p}_i^2 + \frac{1}{2m_\alpha}\sum_\alpha \mathbf{p}_\alpha^2\\ &+\frac{1}{2}\sum_i\sum_{j\neq i}\frac{1}{|\mathbf{r}_i-\mathbf{r}_j|}+\frac{1}{2}\sum_\alpha\sum_{\beta\neq\alpha}\frac{Z_\alpha Z_\beta}{|\mathbf{R}_\alpha-\mathbf{R}_\beta|}-\frac{1}{2}\sum_i\sum_\alpha\frac{Z_\alpha}{|\mathbf{r}_i-\mathbf{R}_\alpha|},\end{aligned} \tag{4.2}$$

where we have adopted atomic units. Latin indices refer to the system's electrons, while Greek indices to the nuclei. From left to right, the above Hamiltonian includes

© The Author(s), under exclusive license to Springer Nature Switzerland AG 2021
A. M. Alvertis, *On Exciton–Vibration and Exciton–Photon Interactions in Organic Semiconductors*, Springer Theses,
https://doi.org/10.1007/978-3-030-85454-6_4

the kinetic energy of the electrons, the kinetic energy of the nuclei, the repulsive electron-electron interaction, the repulsive nucleus-nucleus interaction, and the attractive electron-nucleus interaction.

In principle, diagonalising the Hamiltonian of Eq. (4.2) would yield the full many-body wavefunctions of a system such as an organic molecule. In practice however, obtaining exact solutions to this problem for any realistic system is impossible, since one has to deal with $\sim 10^{23}$ degrees of freedom. Such solutions would also be undesirable, since the resulting many-body wavefunctions would be so complicated that it would be very difficult to extract physical insights.

This complicated many-body problem can be simplified by using the so-called Born-Oppenheimer approximation [2], which we now proceed to discuss, largely following the mathematical formulation of reference [3]. The ratio of the proton to the electron mass is $m_p/m_e = 1836$, therefore atomic nuclei are much heavier and move much slower than electrons, and it can be assumed that electrons react almost instantaneously to their motion. One can thus consider that electrons feel the potential that the nuclei produce when 'frozen' at specific positions $\mathbf{R}$, and hence the electronic Hamiltonian takes the form:

$$\begin{aligned}\mathcal{H}_e = &\frac{1}{2}\sum_i \mathbf{p}_i^2 + \frac{1}{2}\sum_i\sum_{j\neq i}\frac{1}{|\mathbf{r}_i - \mathbf{r}_j|}\\ &-\frac{1}{2}\sum_i\sum_\alpha\frac{Z_\alpha}{|\mathbf{r}_i - \mathbf{R}_\alpha|} + \frac{1}{2}\sum_\alpha\sum_{\beta\neq\alpha}\frac{Z_\alpha Z_\beta}{|\mathbf{R}_\alpha - \mathbf{R}_\beta|}\end{aligned} \tag{4.3}$$

The solutions to the electronic problem $\mathcal{H}_e\,|\psi_n(\mathbf{r};\mathbf{R})\rangle = \varepsilon_n\,|\psi_n(\mathbf{r};\mathbf{R})\rangle$ only depend on the nuclear positions parametrically. These electronic eigenstates form a complete basis of the Hilbert space, so that any function of $\mathbf{r}$ can be written as:

$$|\Psi(\mathbf{r},\mathbf{R})\rangle = \sum_n \chi_n(\mathbf{R})\,|\psi_n(\mathbf{r};\mathbf{R})\rangle\,, \tag{4.4}$$

where the expansion coefficients $\chi_n(\mathbf{R})$ depend on the nuclear coordinate $\mathbf{R}$, and $|\Psi(\mathbf{r};\mathbf{R})\rangle$ is the full many-body wavefunction. Below we will show that in fact, the coefficients $\chi_n(\mathbf{R})$ are simply the nuclear wavefunctions.

By substituting the wavefunction (4.4) into the time-independent Schrödinger equation and using the many-body Hamiltonian (4.2), we obtain:

$$\sum_n\left(\frac{1}{2m_\alpha}\sum_\alpha \mathbf{p}_\alpha^2 + \varepsilon_n(\mathbf{R})\right)\chi_n(\mathbf{R})\,|\psi_n(\mathbf{r};\mathbf{R})\rangle = E\sum_n \chi_n(\mathbf{R})\,|\psi_n(\mathbf{r};\mathbf{R})\rangle\,. \tag{4.5}$$

We now substitute the momentum operator $p_\alpha = -i\nabla_\alpha$ and make use of the following identity for the Laplacian ∇_α^2:

$$\nabla_\alpha^2 \chi_n(\mathbf{R}) \left|\psi_n(\mathbf{r};\mathbf{R})\right\rangle = \left|\psi_n(\mathbf{r};\mathbf{R})\right\rangle \nabla_\alpha^2 \chi_n(\mathbf{R}) + 2\nabla_\alpha \chi_n(\mathbf{R}) \nabla_\alpha \left|\psi_n(\mathbf{r};\mathbf{R})\right\rangle + \chi_n(\mathbf{R}) \nabla_\alpha^2 \left|\psi_n(\mathbf{r};\mathbf{R})\right\rangle , \tag{4.6}$$

which gives:

$$\sum_n \left|\psi_n(\mathbf{r};\mathbf{R})\right\rangle (-\frac{1}{2}\sum_\alpha \frac{1}{m_\alpha}\nabla_\alpha^2 + \varepsilon_n(\mathbf{R}) - E)\chi_n(\mathbf{R}) + \sum_n (-\frac{1}{2}\sum_\alpha \frac{1}{m_\alpha}(2\nabla_\alpha \chi_n(\mathbf{R})\nabla_\alpha \left|\psi_n(\mathbf{r};\mathbf{R})\right\rangle + \chi_n(\mathbf{R})\nabla_\alpha^2 \left|\psi_n(\mathbf{r};\mathbf{R})\right\rangle)) = 0. \tag{4.7}$$

By multiplying the above expression by $\left\langle\psi_m(\mathbf{r};\mathbf{R})\right|$ we obtain:

$$\left(-\frac{1}{2}\sum_\alpha \frac{1}{m_\alpha}\nabla_\alpha^2 + \varepsilon_m(\mathbf{R}) - E\right)\chi_m(\mathbf{R}) + \sum_n [-\frac{1}{2}\sum_\alpha \frac{1}{m_\alpha}(2\nabla_\alpha \chi_n(\mathbf{R}) \left\langle\psi_m(\mathbf{r};\mathbf{R})\right| \nabla_\alpha \left|\psi_n(\mathbf{r};\mathbf{R})\right\rangle + \chi_n(\mathbf{R}) \left\langle\psi_m(\mathbf{r};\mathbf{R})\right| \nabla_\alpha^2 \left|\psi_n(\mathbf{r};\mathbf{R})\right\rangle))] = 0 \tag{4.8}$$

Within the Born-Oppenheimer approximation, the coupling terms, i.e. the sum over n is neglected, and we are finally left with:

$$\left(-\frac{1}{2}\sum_\alpha \frac{1}{m_\alpha}\nabla_\alpha^2 + \varepsilon_m(\mathbf{R})\right)\chi_m(\mathbf{R}) = E_m \chi_m(\mathbf{R}). \tag{4.9}$$

The operator $\mathcal{H}_{\text{vib}} = -\frac{1}{2}\sum_\alpha \frac{1}{m_\alpha}\nabla_\alpha^2 + \varepsilon_m(\mathbf{R})$ is the Hamiltonian for the nuclei moving within the potential created by the electrons, once the electronic state m is populated. Hence $\chi_n(\mathbf{R})$ is the nuclear wavefunction, and we have completely separated the problem into electronic ($\mathcal{H}_e \left|\psi_n(\mathbf{r};\mathbf{R})\right\rangle = \varepsilon_n \left|\psi_n(\mathbf{r};\mathbf{R})\right\rangle$) and nuclear (Eq. 4.9) parts.

We would now like to take a moment and connect the Born-Oppenheimer approximation to the Born approximation of Eq. (3.37). Within the Born approximation, we assumed that the motion of the 'system' could be described for an 'environment' at a given configuration. Here the electrons and the nuclear vibrations represent the system and the environment respectively, and the electronic motion is similarly described for fixed nuclear positions $\mathbf{R}$. However, unlike Eq. (3.37) where the environment is always in its $t = 0$ configuration, here there is a finite feedback on the nuclear motion within the Born-Oppenheimer approximation, through the electronic potential $\varepsilon_m(\mathbf{R})$ of Eq. (4.9).

The Born-Oppenheimer approximation greatly simplifies the solution of the many-body problem, and is used almost universally when it comes to obtaining the time-independent electronic and vibrational properties of material systems. While we will also employ the Born-Oppenheimer approximation to some extent, one needs to be mindful of the limitations to its applicability. In order to evaluate such limitations, let us turn our attention to the largest of the two coupling terms of Eq. (4.8), i.e. the term $\langle \psi_m(\mathbf{r};\mathbf{R})| \nabla_\alpha |\psi_n(\mathbf{r};\mathbf{R})\rangle$. We use:

$$\begin{aligned}\langle \psi_m(\mathbf{r};\mathbf{R})| [\nabla_\alpha, \mathcal{H}_e] |\psi_n(\mathbf{r};\mathbf{R})\rangle &= \langle \psi_m(\mathbf{r};\mathbf{R})| \nabla_\alpha \mathcal{H}_e |\psi_n(\mathbf{r};\mathbf{R})\rangle \\ &\quad - \langle \psi_m(\mathbf{r};\mathbf{R})| \mathcal{H}_e \nabla_\alpha |\psi_n(\mathbf{r};\mathbf{R})\rangle \\ &= [\varepsilon_n(\mathbf{R}) - \varepsilon_m(\mathbf{R})] \langle \psi_m(\mathbf{r};\mathbf{R})| \nabla_\alpha |\psi_n(\mathbf{r};\mathbf{R})\rangle ,\end{aligned} \tag{4.10}$$

in order to write:

$$\langle \psi_m(\mathbf{r};\mathbf{R})| \nabla_\alpha |\psi_n(\mathbf{r};\mathbf{R})\rangle = \frac{\langle \psi_m(\mathbf{r};\mathbf{R})| [\nabla_\alpha, \mathcal{H}_e] |\psi_n(\mathbf{r};\mathbf{R})\rangle}{\varepsilon_n(\mathbf{R}) - \varepsilon_m(\mathbf{R})}. \tag{4.11}$$

From Eq. (4.11) it becomes apparent that the coupling term $\langle \psi_m(\mathbf{r};\mathbf{R})| \nabla_\alpha |\psi_n(\mathbf{r};\mathbf{R})\rangle$ between states m and n can in fact become very large if at a specific nuclear configuration $\mathbf{R}$ the difference in the energy of these states becomes small. Therefore, nuclear motion can induce electronic transitions, leading to a breakdown of the Born-Oppenheimer approximation. This is common occurrence in the excited state dynamics of organic semiconductors, making it necessary to go beyond the Born-Oppenheimer approximation in order to study the time-dependent properties of these systems. We do so by employing the tensor network methods outlined in Chaps. 7 and 8.

4.2 Ground State Properties: Density Functional Theory (DFT)

One of the main challenges associated with the many-body problem, is the accurate treatment of the electron-electron interaction that appears in the Hamiltonian 4.2. Density functional theory (DFT), as formulated by Hohenberg, Kohn, and Sham [4, 5], is in principle an exact way of reducing this many-body problem to the self-consistent solution of a one-particle problem. DFT is arguably the most popular theory for calculating ground state properties from first principles, and may be used to obtain the total energy of a many-body system, as well as structural and vibrational properties. While DFT is a ground state theory, it can serve as a basis for the excited state calculations that we perform using the PPP and GW-BSE approaches described in Sects. 4.3 and 4.4 respectively. Our following presentation of the basic ideas of DFT is based on Ref. [6].

4.2.1 Fundamentals

The central idea within DFT is to express the ground state energy as a functional of the electronic charge density $\rho(\mathbf{r})$, rather than the many-body wavefunction Ψ. For an interacting many-electron system in an external potential $v(\mathbf{r})$, the so-called Hohenberg and Kohn theorems [4] state that:

1. The energy of the system is $E[\rho] = \int v(\mathbf{r})\rho(\mathbf{r})d\mathbf{r} + F[\rho]$, where $F[\rho]$ a functional of the density which is independent of the external potential.
2. $E[\rho]$ is minimised for the correct physical density, which satisfies the constraint: $N = \int \rho(\mathbf{r})d\mathbf{r}$, where N the total number of electrons.

There is no general form for the functional $F[\rho]$, which contains both the kinetic energy of the electrons, as well as the electron-electron interaction. However, it is possible to obtain $F[\rho]$ in some limiting cases. Consider for example the case of a non-interacting system of electrons with density $\rho(\mathbf{r})$. If $\varphi_i(\mathbf{r})$ the single-particle orbital representing electron i, then:

$$\rho(\mathbf{r}) = \sum_i \varphi_i^*(\mathbf{r})\varphi_i(\mathbf{r}), \tag{4.12}$$

and the associated kinetic energy:

$$T_{\mathrm{NI}}[\rho] = -\frac{1}{2}\sum_i \int \varphi_i^*(\mathbf{r})\nabla^2\varphi_i(\mathbf{r})d\mathbf{r}, \tag{4.13}$$

where the index i runs over all occupied states.

If we now allow for Coulomb interactions between electrons, but ignore the effects of exchange and correlation, then the interaction energy is given as a Hartree term:

$$E_{\mathrm{Hartree}} = \frac{1}{2}\int \frac{\rho(\mathbf{r})\rho(\mathbf{r}')}{|\mathbf{r}-\mathbf{r}'|}d\mathbf{r}d\mathbf{r}'. \tag{4.14}$$

It is therefore intuitive to write the total energy of an interacting system as:

$$E[\rho] = \int v(\mathbf{r})\rho(\mathbf{r})d\mathbf{r} + T_{\mathrm{NI}}[\rho] + \frac{1}{2}\int \frac{\rho(\mathbf{r})\rho(\mathbf{r}')}{|\mathbf{r}-\mathbf{r}'|}d\mathbf{r}d\mathbf{r}' + E_{\mathrm{XC}}[\rho], \tag{4.15}$$

where the term $E_{\mathrm{XC}}[\rho]$ is called the exchange-correlation functional and contains all unknown terms. These include the electronic interaction beyond the Hartree approximation, as well as part of the kinetic energy of interacting electrons, since $T_{\mathrm{NI}}[\rho]$ refers to the non-interacting system.

What Kohn and Sham showed [5], was that the electron density and total energy of an interacting system, may be obtained by solving an effective system with the same number of non-interacting electrons N, which move in a potential $v'(\mathbf{r})$. The eigenstates and eigenvalues of this non-interacting system are obtained by solving:

$$\left[-\frac{1}{2}\nabla^2 + v'(\mathbf{r})\right]\varphi_i'(\mathbf{r}) = \varepsilon_i'\varphi_i'(\mathbf{r}), \tag{4.16}$$

and hence the electronic density is $\rho'(\mathbf{r}) = \sum_{i=1}^{N} |\varphi_i'(\mathbf{r})|^2$. This density ρ' is used as input to evaluate the total energy $E[\rho']$ of the original, interacting system. If one minimises $E[\rho']$ with respect to ρ', then from the second Hohenberg-Kohn theorem, this is also the density which correctly describes the interacting system. The minimisation of $E[\rho']$ leads to the following relation between the potential within which the electrons of the non-interacting system move in, and the interacting system:

$$v'(\mathbf{r}) = v(\mathbf{r}) + \int \frac{\rho(\mathbf{r})\rho(\mathbf{r}')}{|\mathbf{r}-\mathbf{r}'|} d\mathbf{r}d\mathbf{r}' + \frac{\delta E_{\mathrm{XC}}[\rho]}{\delta\rho}. \tag{4.17}$$

Using the potential (4.17) in Eq. (4.16), yields the wavefunctions and hence the electronic density of the non-interacting system. The resulting density is also that of the interacting system, allowing us to obtain its total energy and other observables. The set of Eq. (4.16), which are referred to as the Kohn-Sham equations, can be solved self-consistently; the potential $v'(\mathbf{r})$ depends on the density, and we use it to obtain the $\varphi_i'(\mathbf{r})$ wavefunctions, which in turn may be used to obtain an improved estimate of the density and hence $v'(\mathbf{r})$, and so on.

The DFT formalism within the Kohn-Sham framework provides a tremendous simplification of the many-electron problem. Instead of being required to solved a $3N$-dimensional problem, one is faced with the much simpler task of solving a set of independent Kohn-Sham Equations (4.16).

4.2.2 *Approximations Within DFT*

4.2.2.1 Choice of Functional

Despite the fact that DFT is in principle exact, the functional E_{XC} is unknown, and a number of approximations have been used in order to give it a concrete form. Such a common approximation is to write:

$$E_{\mathrm{XC}} = \int \rho(\mathbf{r})\varepsilon_{\mathrm{XC}}(\mathbf{r})d\mathbf{r}, \tag{4.18}$$

where $\varepsilon_{\mathrm{XC}}(\mathbf{r})$ an exchange-correlation energy density. Within the so-called local density approximation (LDA), one assumes that $\varepsilon_{\mathrm{XC}}$ is only a function of the local density: $\varepsilon_{\mathrm{XC}} = \varepsilon_{\mathrm{XC}}(\rho(\mathbf{r}))$, where $\rho(\mathbf{r})$ the density of the uniform electron gas. Within the generalised gradient approximation (GGA) the dependence on $\nabla\rho(\mathbf{r})$ is also included in the exchange-correlation energy density: $\varepsilon_{XC} = \varepsilon_{XC}(\rho(\mathbf{r}), \nabla\rho(\mathbf{r}))$. In this thesis, we make extensive use of the GGA when studying crystalline systems in Chaps. 5 and 6, using the functional designed by Perdew, Burke and Ernzerhof (PBE) in [7]. While it is possible to improve upon the accuracy of DFT by using hybrid functionals (see below), this can quickly become prohibitively expensive for extended systems.

One of the shortcomings of LDA and GGA is that they do not give the correct long-range $\frac{1}{r}$ asymptotic behaviour for the exchange term. On the other hand, the Hartree-Fock approximation does, giving the exchange energy:

$$E_X[\rho] = -\frac{1}{2}\sum_{i,j}\int d\mathbf{r}d\mathbf{r}'\frac{\varphi_i^*(\mathbf{r})\varphi_j^*(\mathbf{r}')\varphi_j^*(\mathbf{r})\varphi_i^*(\mathbf{r}')}{|\mathbf{r}-\mathbf{r}'|} \tag{4.19}$$

The associated potential entering the Kohn-Sham equations is then:

$$v_X[\rho](\mathbf{r}) = \frac{\delta E_X[\rho]}{\delta\rho(\mathbf{r})} \tag{4.20}$$

While LDA and GGA tend to underestimate the band gaps of condensed matter systems, Hartree-Fock tends to overestimate them [8]. A way of improving the agreement between DFT results and experiment is the use of hybrid functionals, which incorporate a fraction of Hartree-Fock exchange, along with the exchange-correlation term e.g. from GGA. The precise balance between the exchange terms is controlled through a number of parameters, which are typically chosen empirically. One of the most widely used functionals is the Becke, 3-parameter, Lee-Yang-Parr (B3LYP) functional [9], which we will also make use of when studying the molecular systems of Chaps. 7–9.

4.2.2.2 Choice of Basis

Another approximation which is necessary when performing DFT calculations, is that of choosing a finite basis in which to express the calculated Kohn-Sham wavefunctions. There are two kinds of basis sets we will consider: planes waves, for the case of periodic solids, and atomic orbitals for isolated molecules.

Let us start from the case of an infinite periodic system. According to Bloch's theorem, the wavefunction of an electron with wavevector $\mathbf{k}$ can be written as:

$$\Psi_\mathbf{k}(\mathbf{r}) = e^{i\mathbf{k}\cdot\mathbf{r}}u_\mathbf{k}(\mathbf{r}), \tag{4.21}$$

where the function $u_{\mathbf{k}}(\mathbf{r})$ has the periodicity of the lattice:

$$u_{\mathbf{k}}(\mathbf{r}) = u_{\mathbf{k}}(\mathbf{r} + \mathbf{R}_n), \tag{4.22}$$

with $\mathbf{R}_n$ a lattice vector. Therefore $u_{\mathbf{k}}(\mathbf{r})$ may be written as a discrete Fourier sum:

$$u_{\mathbf{k}}(\mathbf{r}) = \sum_{\mathbf{G}} b_{\mathbf{k}}(\mathbf{G}) e^{i\mathbf{G}\cdot\mathbf{r}}. \tag{4.23}$$

The vectors $\mathbf{G}$ are reciprocal lattice vectors. The electronic wavefunction is now written as:

$$\Psi_{\mathbf{k}}(\mathbf{r}) = e^{i\mathbf{k}\cdot\mathbf{r}} \sum_{\mathbf{G}} b_{\mathbf{k}}(\mathbf{G}) e^{i\mathbf{G}\cdot\mathbf{r}}, \tag{4.24}$$

i.e. as an infinite sum over plane waves. In practice however, one can only include a finite number of terms in order to describe a wavefunction when performing a computation. The maximal value of the $\mathbf{G}$ vector which is included in the sum is commonly expressed in terms of a kinetic energy cutoff: $|\mathbf{G}|^2 \leq E_c$. The cutoff energy E_c is a parameter in respect with which one needs to converge the results of a simulation.

A different kind of basis which is commonly employed when studying non-periodic systems is that of atomic orbitals. This essentially includes forming linear combinations of functions (e.g. Gaussians) that represent atomic orbitals and are centred around the different atomic nuclei. Unlike plane wave basis sets for which one can simply increase E_c in order to obtain more accurate results, there is no systematic way to achieve the same in the case of atomic orbitals. However, for molecular systems these basis sets can be very efficient, as the number of basis functions is usually very small compared to plane waves. The efficiency of atomic orbital basis sets makes it possible to combine them with the use of hybrid functionals, while maintaining a reasonable computational cost. Indeed for the isolated molecular systems of Chaps. 7–9 we use the B3LYP hybrid functional alongside the atomic orbitals basis set cc-pVDZ, while for the periodic solids of Chaps. 5 and 6 we resort to plane waves coupled to the GGA.

4.2.2.3 Pseudopotentials

The behaviour of outer valence electrons largely determines the properties of materials, as they are the ones that determine effective interatomic interactions. In contrast to that, core electrons are tightly bound to the atomic nucleus, and the effect of other atoms on them is negligible. Therefore, the behaviour of core electrons in a material system resembles that of core electrons in isolated atoms.

The above observation forms the basic idea behind the theory of pseudopotentials, developed by Philips [10], Cohen and Heine [11]. Core electrons behave very similarly between material systems and their dominant effect is to weaken the interaction of the valence electrons with the atomic nucleus. It is therefore intuitive to replace their combined effect with an effective potential, which eliminates the core electrons from the problem, but leaves the properties of the valence electrons largely unaltered compared to the all-electron case. This provides great efficiency in computation, allowing for using much simpler basis sets. While pseudopotentials are generated from first principles, we will use existing ones throughout this thesis. These are generally very accurate and transferable, satisfying constraints such as the norm conservation of the atomic orbitals.

4.2.3 Applications

DFT is one of the most powerful and versatile approaches for obtaining ground state properties of materials. Its applications are numerous, ranging from the study of magnetic properties to the charge dynamics of biological molecules. Here we will briefly outline the application of DFT towards two problems that are relevant for this thesis: performing structural optimisations and computing the vibrational modes of materials.

4.2.3.1 Geometry Optimisations

The structural properties of a system can be obtained by minimising its total ground state energy E_{GS}, in respect with the atomic positions. This can be done by using a number of different minimisation algorithms that depend on computing the gradient of the total energy, i.e. the forces on the atoms:

$$\mathbf{F}_{\alpha} = -\frac{\partial E_{\mathrm{GS}}}{\partial \mathbf{R}_{\alpha}}. \tag{4.25}$$

Depending on the energy landscape of the system at hand, one might employ simpler algorithms that follow the steepest descent of the total energy, or more complicated ones based on conjugate gradients [12].

Finding the relaxed geometry of a structure is very important in order to obtain accurate excited state properties. It is also crucial for the calculation of molecular and crystal vibrations. Therefore, in every case when we perform an electronic structure or vibrational calculation throughout this thesis, a geometry optimisation of the structure is performed prior to that, unless explicitly stated otherwise. Computational details for optimising the structure of the different studied systems are given in the respective chapters.

4.2.3.2 Calculation of Vibrational Modes

Let us now discuss how DFT may be used to obtain the vibrational properties of an infinite periodic system. The discussion naturally extends to finite molecular systems when we consider the limit of an infinite lattice constant and hence we limit ourselves to the Γ point $\mathbf{q} = \mathbf{0}$ of the first Brillouin zone. By denoting the different unit cells with lattice vectors $\mathbf{R}_p$ by p, and the atoms within them by α, the nuclear Hamiltonian of Eq. (4.9) may be rewritten for the case of a material exhibiting oscillations with small displacements $\tau_{p\alpha i}$ of its atoms in the spatial direction i around equilibrium, by expanding the potential energy term to second order in atomic displacement:

$$\mathcal{H}_{\text{vib}} = -\frac{1}{2}\sum_{\mathbf{R}_p,\alpha}\frac{1}{m_\alpha}\nabla^2_{p\alpha} + \frac{1}{2}\sum_{\mathbf{R}_p,\alpha,i}\sum_{\mathbf{R}_{p'},\alpha',i'}\frac{\partial^2 E_{\text{GS}}}{\partial\tau_{p\alpha i}\partial\tau_{p'\alpha'i'}}\tau_{p\alpha i}\tau_{p'\alpha'i'} \tag{4.26}$$

where ε_m has been substituted with the total electronic ground state energy E_{GS} as obtained after a geometry optimisation. Since we have expanded E_{GS} around the equilibrium position, the linear term of the Taylor expansion vanishes, and the second-order term is the largest one which survives. We ignore higher-order terms within the harmonic approximation, and we have also set the total energy of the system $E_{\text{GS}}(\mathbf{R} = \mathbf{R}_{\text{eq}})$ at equilibrium to zero. The matrix:

$$A_{i\alpha,i'\alpha'}(\mathbf{R}_p, \mathbf{R}_{p'}) = \frac{\partial^2 E_{\text{GS}}}{\partial\tau_{p\alpha i}\partial\tau_{p'\alpha'i'}}, \tag{4.27}$$

contains the force constants of the harmonic restoring forces on an atom, due to its displacement relative to another atom. Then for atom α the classical equation of motion yields:

$$m_\alpha \ddot{\tau}_{p\alpha i} = -\sum_{\mathbf{R}_{p'},\alpha',i'} A_{i\alpha,i'\alpha'}(\mathbf{R}_p, \mathbf{R}_{p'})\tau_{p'\alpha'i'}. \tag{4.28}$$

Vibrational normal modes of the system represent collective motions, within which every atom moves with the same frequency ω. We therefore look for solutions of the form:

$$\tau_{p\alpha i}(t) = \frac{C_{p\alpha i}}{m_\alpha}e^{-i\omega t}, \tag{4.29}$$

where $C_{p\alpha i}$ the amplitude of the vibration. We can exploit the periodicity of the lattice and use a Bloch-like form for the amplitudes:

$$C_{p\alpha i} = C_{\alpha i}e^{i\mathbf{q}\cdot(\mathbf{R}_p+\mathbf{r}_\alpha)}, \tag{4.30}$$

where $\mathbf{q}$ a reciprocal lattice vector, and $\mathbf{r}_\alpha$ the basis vector giving the position of atom α within unit cell p. We thus get the following eigenvalue problem:

$$\omega^2 C_{\alpha i} = \sum_{\alpha',i'} D_{i\alpha,i'\alpha'}(\mathbf{q}) C_{\alpha' i'}, \tag{4.31}$$

where:

$$D_{i\alpha,i'\alpha'}(\mathbf{q}) = e^{-i\mathbf{q}\cdot(\mathbf{r}_\alpha - \mathbf{r}_{\alpha'})} \sum_p \frac{A_{i\alpha,i'\alpha'}(\mathbf{R}_p, \mathbf{R}_{p'})}{\sqrt{m_\alpha m_{\alpha'}}} e^{-i\mathbf{q}\cdot(\mathbf{R}_p - \mathbf{R}_{p'})} \tag{4.32}$$

the so-called dynamical matrix. In the context of isolated molecular systems (single unit cell and $\mathbf{q} \to 0$), this is often referred to as the Hessian matrix.

The eigenvectors of Eq. (4.31) are the vibrational modes of the system, which are referred to as phonons in the case of periodic solids. For isolated molecular systems we simply refer to vibrational modes, which are sometimes also called vibrons in the literature. The main prerequisite for a calculation of the vibrational structure of a material is therefore the dynamical matrix.

It is the matrix of force constants (4.27) that DFT is often used for calculating. There exist two main methods for obtaining the matrix of force constants (4.27) and hence the vibrational modes of a system: density functional perturbation theory (DFPT) [13] and finite differences (sometimes referred to as frozen phonon) approaches [14]. While both have their advantages and disadvantages, the strong exciton-vibration interactions that occur in organic semiconductors make finite differences methods more appropriate for our purposes, as we argue in Chap. 6.

Within the finite differences method, we calculate the matrix of force constants as described in [3]. We impose a small displacement on every atom in the studied system, one by one. The displacement $\delta\tau_{p,\alpha,i}$ of atom α in unit cell p and in the spatial direction i leads to a small but finite force $\delta F^{p',\alpha',i'}_{p,\alpha,i}$ due to another atom α', which is found in some unit cell p' and direction i'. According to the Hellman-Feynman theorem [15], this force is equal to:

$$\delta F^{p',\alpha',i'}_{p,\alpha,i} = \frac{\partial E_{\text{GS}}}{\partial \tau_{p',\alpha',i'}}. \tag{4.33}$$

Hence one obtains the matrix of force constants as:

$$A_{i\alpha,i'\alpha'}(\mathbf{R}_p, \mathbf{R}_{p'}) = \frac{\partial^2 E_{\text{GS}}}{\partial \tau_{p\alpha i} \partial \tau_{p'\alpha' i'}} = \frac{\partial F^{p',\alpha',i'}_{p,\alpha,i}}{\partial \tau_{p\alpha i}} \approx \frac{\delta F^{p',\alpha',i'}_{p,\alpha,i}}{\delta \tau_{p\alpha i}} \tag{4.34}$$

4.3 Excited State Properties: Model Hamiltonians and the PPP Method

We now turn our attention to the excited states of material systems. While one can obtain excited state properties from first principles, for example within the so-called GW approximation outlined in Sect. 4.4, it is in some cases useful to rely on model

Hamiltonians that reduce the complexity of the problem at hand. In this thesis, we make use of the model Hamiltonian proposed by Parr et al. [16, 17], referred to as the PPP Hamiltonian, in order to obtain the exciton properties of the molecular dimer systems studied in Chaps. 7 and 9. We will see later that in these cases the PPP approach gives excellent agreement to experiment, while only requiring very modest computational resources. We now proceed to discuss the basic steps for the derivation of such model Hamiltonians, closely following Ref. [18].

As discussed in Chap. 2, the excited state properties of organic semiconductors are largely determined by weakly bound π molecular orbitals that result from the hybridisation of the $2p_z$ carbon atomic orbitals. The electrons occupying the rest of the carbon atomic orbitals participate in strong σ bonds, and their effect is mostly to renormalise the ion-ion and π-electron-ion interactions, as well as to screen the interaction between π electrons. It is therefore a reasonable first approximation to explicitly treat π electrons of organic semiconductors, and only consider the effects of σ electrons implicitly. If we therefore start from the Born-Oppenheimer electronic Hamiltonian of Eq. (4.3), we can replace the electron-ion interaction with a pseudopotential $V_p(\mathbf{r}_i, \mathbf{R})$, which not only incorporates the effect of core electrons as the pseudopotentials discussed in Sect. 4.2.2, but also that of σ electrons. Furthermore, the electron-electron interaction term is replaced by an effective potential $V_{\text{eff}}^{e-e}(\mathbf{r})$, which only depends on the coordinates of π electrons, and incorporates the screening because of core and σ electrons via a background dielectric constant ε:

$$V_{\text{eff}}^{e-e}(\mathbf{r}) = \sum_{j>i} \frac{1}{\varepsilon|\mathbf{r}_i - \mathbf{r}_j|} = \sum_{j>i} V_{\text{eff}}^{e-e}(\mathbf{r}_i - \mathbf{r}_j). \tag{4.35}$$

Therefore, working in the basis $\{\varphi\}$ of $2p_z$ electrons centred around the carbon atoms of the studied system, the π electron Hamiltonian in second quantisation may be written as:

$$\mathcal{H}_{\pi e} = -\sum_{l,m,\sigma} t_{l,m} c_{l,\sigma}^{\dagger} c_{m,\sigma} + \frac{1}{2} \sum_{ij,kl,\sigma\sigma'} V_{ij,kl} c_{i,\sigma}^{\dagger} c_{j,\sigma'}^{\dagger} c_{l,\sigma'} c_{k,\sigma}, \tag{4.36}$$

where the operator $c_{l,\sigma}^{\dagger}$ creates an electron of spin σ on atom l, and $c_{l,\sigma}$ destroys it. Here we have the one-electron term:

$$t_{l,m} = -\int d\mathbf{r} \varphi_l^*(\mathbf{r}) \left[\frac{\mathbf{p}^2}{2} + V_p(\mathbf{r}, \mathbf{R}) \right] \varphi_m(\mathbf{r}), \tag{4.37}$$

and the two-electron term:

$$V_{ij,kl} = \int d\mathbf{r} \int d\mathbf{r}' \varphi_i^*(\mathbf{r}) \varphi_j^*(\mathbf{r}') V_{\text{eff}}^{e-e}(\mathbf{r} - \mathbf{r}') \varphi_k(\mathbf{r}) \varphi_l(\mathbf{r}'). \tag{4.38}$$

To make further progress we expand the pseudopotential around the equilibrium nuclear configuration $\mathbf{R}_o$ of the studied structure:

$$V_p(\mathbf{r}, \mathbf{R}) = V_p(\mathbf{r}, \mathbf{R}_o) + \sum_l \frac{\partial V_p}{\partial \mathbf{R}_l}(\mathbf{r}, \mathbf{R}_o) \cdot \mathbf{u}_l + \mathcal{O}(\mathbf{u}_l^2), \tag{4.39}$$

with $\mathbf{u}_l = \mathbf{R}_l - \mathbf{R}_{l,o}$ a small displacement of ion l from its equilibrium position. By using this expansion, and ignoring higher order terms, the one-electron term of the Hamiltonian (4.36) can be written as:

$$\sum_{m,n,\sigma} t_{m,n} c^\dagger_{m,\sigma} c_{n,\sigma} = \sum_{m,\sigma} \varepsilon_m c^\dagger_{m,\sigma} c_{n,\sigma} + \sum_{m \neq n,\sigma} t^o_{m,n} c^\dagger_{m,\sigma} c_{n,\sigma} + \sum_{m,n,l,\sigma} \mathbf{a}_{m,n,l} \cdot \mathbf{u}_l c^\dagger_{m,\sigma} c_{n,\sigma}, \tag{4.40}$$

where we have introduced the on-site energies:

$$\varepsilon_m = \int d\mathbf{r} \varphi^*_m(\mathbf{r}) \left[\frac{\mathbf{p}^2}{2} + V_p(\mathbf{r}, \mathbf{R}_o) \right] \varphi_m(\mathbf{r}), \tag{4.41}$$

the bare transfer integrals:

$$t^o_{m,n} = \int d\mathbf{r} \varphi^*_m(\mathbf{r}) \left[\frac{\mathbf{p}^2}{2} + V_p(\mathbf{r}, \mathbf{R}_o) \right] \varphi_n(\mathbf{r}), \tag{4.42}$$

and the electron-phonon coupling term:

$$\mathbf{a}_{m,n,l} = \int d\mathbf{r} \varphi^*_m(\mathbf{r}) \frac{\partial V_p}{\partial \mathbf{R}_l} \varphi_n(\mathbf{r}). \tag{4.43}$$

For the moment we will ignore the electron-phonon term, and focus on electron-electron interactions, in order to derive a Hamiltonian for the excited states of a π electron system at the equilibrium geometry of a system. We will however bring electron-phonon (or rather exciton-phonon) terms back, when we study the dynamics of charge transfer and singlet fission in a tetracene dimer, in Chaps. 7 and 9 respectively.

Let us now examine the terms that appear due to the electron-electron interaction in the Hamiltonian (4.36). The diagonal term:

$$U = V_{ii,ii} = \int d\mathbf{r} \int d\mathbf{r}' |\varphi_i(\mathbf{r})|^2 V^{e-e}_{\text{eff}}(\mathbf{r} - \mathbf{r}') |\varphi_i(\mathbf{r}')|^2, \tag{4.44}$$

describes the scattering of electrons centred around the same site i, whereas the diagonal term:

$$V_l = V_{i,i+l,i,i+l} = \int d\mathbf{r} \int d\mathbf{r}' |\varphi_i(\mathbf{r})|^2 V^{e-e}_{\text{eff}}(\mathbf{r} - \mathbf{r}') |\varphi_{i+l}(\mathbf{r}')|^2, \tag{4.45}$$

refers to scattering between two electrons located on sites i and $i+l$.The magnitude of the terms V_l depends on how rapidly the Coulomb interaction decays with distance within a given material. We will return to this point, when we choose a specific form for V_l below.

While the diagonal terms appearing in (4.44) and (4.45) do not depend on the overlap of the electronic wavefunctions centred around the different carbon atoms (they only depend on the electronic density), general off-diagonal terms do, as for example the term:

$$V_{ii,i,i+1} = \int d\mathbf{r} \int d\mathbf{r}' |\varphi_i(\mathbf{r})|^2 V_{\text{eff}}^{e-e}(\mathbf{r}-\mathbf{r}')\varphi_i^*(\mathbf{r}')\varphi_{i+1}(\mathbf{r}'). \tag{4.46}$$

The overlap of these site-centred orbitals is small, and we neglect any off-diagonal terms such as the one of Eq. (4.46), within the so-called zero differential overlap (ZDO) approximation, which has been commonly employed for similar systems [19].

Following all of the aforementioned approximations, in addition to only including the transfer integrals (4.42) between nearest neighbours, and setting the reference energy such that the on-site energies (4.41) vanish, we are left with the PPP Hamiltonian for π electrons, proposed by Pariser et al. in [16, 17]:

$$\mathcal{H}_{PPP} = -\sum_{l,\sigma} t^o_{l,l+1}(c^\dagger_{l,\sigma}c_{l+1,\sigma} + c^\dagger_{l+1,\sigma}c_{l,\sigma}) + U\sum_l n_{l\uparrow}n_{l\downarrow} + \sum_{m,l\leq 1} V_l n_m n_{m+l}, \tag{4.47}$$

with n_l being the density of electrons $|\varphi_l(\mathbf{r})|^2$ on site l. For performing a computation, one needs to specify the parameters U and V. For $V=0$, we recover the famous Hubbard model [20]. For the long-range electron-electron interaction, we use the Ohno parametrisation [21]:

$$V_l = \frac{U}{[1+(\frac{|\mathbf{r}_i-\mathbf{r}_j|}{\alpha_o})^2]^{1/2}}, \quad i-j=l, \tag{4.48}$$

where the Ohno parameter α_o determines the distance over which the Coulomb interaction decays.

4.4 Excited State Properties: Many-Body Perturbation Theory

For many-electron systems, the Coulomb interaction between a given electron and hole is screened by the large number of charge carriers. Moreover, the presence of an additional electron or hole leads to the reorganisation of the electronic cloud of the material, inducing a new potential that acts back on the additional particle.

The accurate description of such effects is particularly important when it comes to extended solid state systems, however DFT fails to capture them. Instead, they can be treated within many-body perturbation theory (MBPT). This section provides an introduction to basic elements of MBPT, including the formalism of Green's functions and the concept of self-energy. We apply these elements towards deriving the so-called GW approximation for quasiparticle properties, and then go beyond that in order to obtain the Bethe-Salpeter equation for excitons. We refer the reader interested in a more detailed discussion of these subjects to the review article by Onida et al. [8].

4.4.1 Green's Functions and the Dyson Equation

Let us begin the discussion by introducing the field operators for a many-electron system. In a plane wave basis, the following operator destroys an electron at position $\mathbf{r}$ with spin σ:

$$\psi_\sigma(\mathbf{r}) = \frac{1}{\sqrt{V}} \sum_{\mathbf{k}} c_{\mathbf{k},\sigma} e^{i\mathbf{k}\cdot\mathbf{r}}, \tag{4.49}$$

where similar to the previous section, the electron annihilation operator $c_{\mathbf{k},\sigma}$ destroys an electron with wavevector $\mathbf{k}$ and spin σ. For the example case of an interacting electron gas in the presence of a positive background charge distribution $n(\mathbf{r})$, field operators allow us to write the Hamiltonian in the form [22]:

$$\begin{aligned}
\mathcal{H} = &\sum_\sigma \int d\mathbf{r} \psi_\sigma^\dagger(\mathbf{r}) \Big(-\frac{\nabla^2}{2} \Big) \psi_\sigma(\mathbf{r}) \\
&+ \frac{1}{2} \sum_{\sigma,\sigma'} \int d\mathbf{r} d\mathbf{r}' \psi_\sigma^\dagger(\mathbf{r}) \psi_{\sigma'}^\dagger(\mathbf{r}') \frac{1}{|\mathbf{r}-\mathbf{r}'|} \psi_{\sigma'}(\mathbf{r}') \psi_\sigma(\mathbf{r}) \\
&- \sum_\sigma \int d\mathbf{r} d\mathbf{r}' \psi_\sigma^\dagger(\mathbf{r}) \frac{n(\mathbf{r}')}{|\mathbf{r}-\mathbf{r}'|} \psi_\sigma(\mathbf{r}) + \frac{1}{2} \int d\mathbf{r} d\mathbf{r}' \frac{n(\mathbf{r}) n(\mathbf{r}')}{|\mathbf{r}-\mathbf{r}'|},
\end{aligned} \tag{4.50}$$

where the first term represents the kinetic energy of the electron, followed by the electron-electron repulsion, the attractive electron-nucleus interaction, and finally the repulsive nucleus-nucleus term.

In order to study the properties of a many-body system as the one described by the Hamiltonian (4.50), it is useful to define the retarded Green's function for electrons:

$$G^r_{\sigma\sigma'}(\mathbf{r}t, \mathbf{r}'t') = -i\Theta(t-t') \langle [\psi_\sigma(\mathbf{r},t), \psi_{\sigma'}^\dagger(\mathbf{r}',t')]_+ \rangle. \tag{4.51}$$

Here the expectation value is calculated according to Eq. (3.8), where the density matrix ρ is computed at an initial time, such as $t = -\infty$. The symbol $[...]_+$ denotes an anticommutator, with the field operators satisfying the relations:

$$
\begin{aligned}
[\psi_\sigma(\mathbf{r}), \psi_{\sigma'}(\mathbf{r}')]_+ = [\psi_\sigma^\dagger(\mathbf{r}), \psi_{\sigma'}^\dagger(\mathbf{r}')]_+ = 0,\\
[\psi_\sigma(\mathbf{r}), \psi_{\sigma'}^\dagger(\mathbf{r}')]_+ = \delta(\mathbf{r}-\mathbf{r}')\delta_{\sigma\sigma'}.
\end{aligned} \tag{4.52}
$$

The step function $\Theta(t-t')$ appearing in Eq. (4.51) ensures that the Green's function is retarded, i.e. $G_r = 0$ for $t' > t$.

By using a few algebraic manipulations, it is possible to show that this quantum mechanical Green's function obeys an equation of motion characterised by a delta function inhomogeneity, similar to its classical equivalent:

$$
\begin{aligned}
i\frac{\partial}{\partial t}G^r_{\sigma\sigma'}(\mathbf{r}t, \mathbf{r}'t') &= \delta(t-t')\delta(\mathbf{r}-\mathbf{r}')\delta_{\sigma\sigma'}\\
&\quad - i\Theta(t-t')\langle[[\psi_\sigma(\mathbf{r},t), \mathcal{H}(t)], \psi^\dagger_{\sigma'}(\mathbf{r}',t')]_+\rangle
\end{aligned} \tag{4.53}
$$

Following Ref. [22], for the interacting electron gas described by the Hamiltonian (4.50), this equation of motion takes the form:

$$
\begin{aligned}
\left(i\frac{\partial}{\partial t} - \frac{\nabla^2}{2}\right)G^r_{\sigma\sigma'}(\mathbf{r}t, \mathbf{r}'t') &= \delta(t-t')\delta(\mathbf{r}-\mathbf{r}')\delta_{\sigma\sigma'}\\
&\quad + \sum_{\sigma''}\int d\mathbf{r}''dt''\Sigma^r_{\sigma\sigma''}(\mathbf{r}t, \mathbf{r}''t'')G^r_{\sigma''\sigma'}(\mathbf{r}''t'', \mathbf{r}'t'),
\end{aligned} \tag{4.54}
$$

where we have introduced the *self-energy* Σ^r, which is the term appearing as a result of many-body interactions. For the specific case of the interacting electron gas in the presence of a background charge density, we find:

$$
\begin{aligned}
\Sigma^r_{\sigma\sigma''}(\mathbf{r}t, \mathbf{r}''t'') &= \delta(t-t')\{\delta(\mathbf{r}-\mathbf{r}')\delta_{\sigma\sigma''}\cdot\int d\mathbf{r}_1 V(\mathbf{r}-\mathbf{r}_1)\\
&\quad \left[\sum_{\sigma_1}\langle\psi^\dagger_{\sigma_1}(\mathbf{r}_1,t)\psi_{\sigma_1}(\mathbf{r}_1,t)\rangle - n(\mathbf{r}_1)\right]\\
&\quad - V(\mathbf{r}-\mathbf{r}'')\langle\psi^\dagger_{\sigma''}(\mathbf{r}'',t)\psi_\sigma(\mathbf{r},t)\rangle.
\end{aligned} \tag{4.55}
$$

Equation (4.54) is the so-called *Dyson equation* for the retarded Green's function. By performing a Fourier transform with respect to spatial and temporal coordinates, we find:

$$\left(\omega - \frac{k^2}{2}\right) G^r_{\sigma\sigma}(\mathbf{k}, \omega) = 1 + \Sigma^r_{\sigma\sigma}(\mathbf{k}, \omega) G^r_{\sigma\sigma}(\mathbf{k}, \omega), \tag{4.56}$$

which can be rewritten as:

$$G^r_{\sigma\sigma}(\mathbf{k}, \omega) = \frac{1}{\omega + i\delta - \varepsilon_k - \Sigma^r_{\sigma\sigma}(\mathbf{k}, \omega)}, \tag{4.57}$$

with $\delta \to 0^+$. The form of Eq. (4.57) for the Green's function highlights the fact that the energy ε_k of a free particle is replaced by $\varepsilon_k + \Sigma^r_{\sigma\sigma}(\mathbf{k}, \omega)$ as a consequence of the many-body interactions, which is why Σ^r is referred to as the self-energy. Since generally the self-energy assumes complex values, its real part results in a shift of the single-particle energies due to the interactions, while its imaginary part leads to a finite broadening of their values.

While this still allows one to think in terms of single particles and their energies, these now contain the effects of all other particles and are referred to as *quasiparticles*. As it has been shown in Ref. [23], the Dyson equation can be cast into the quasiparticle picture:

$$[\mathcal{H}_o + \Sigma(E^{\text{QP}})] \left|\psi^{\text{QP}}\right\rangle = E^{\text{QP}} \left|\psi^{\text{QP}}\right\rangle, \tag{4.58}$$

which is a one-particle Schrödinger-like equation that includes many-body effects through the self-energy.

4.4.2 Quasiparticles: The *GW* Approximation

For obtaining the properties of quasi-electrons and quasi-holes one needs to solve the Dyson Eq. (4.58). However, this requires explicit knowledge of the self-energy operator. Hedin [24] generated a perturbation series of the self-energy in terms of the screened Coulomb interaction W, allowing him to derive a self-consistent scheme for obtaining the self-energy, through a set of coupled equations:

$$\Sigma(1, 2) = i \int G(1, 3)\Gamma(3, 2, 4)W(4, 1^+)d(34), \tag{4.59}$$

$$W(1, 2) = v(1, 2) + \int v(1, 3)P(3, 4)W(4, 2)d(34), \tag{4.60}$$

$$P(1, 2) = -i \int G(1, 3)G(4, 1^+)\Gamma(3, 4, 2)d(34), \tag{4.61}$$

$$\Gamma(1, 2, 3) = \delta(1, 2)\delta(1, 3) + \int \frac{\delta\Sigma(1, 2)}{\delta G(4, 5)} G(4, 6)G(7, 5)\Gamma(6, 7, 3)d(4567), \tag{4.62}$$

where v is the bare Coulomb interaction (as opposed to the screened Coulomb interaction W). Γ is the so-called vertex function, which contains information on electron-hole interactions, since it is determined by the change in the self-energy upon excitation. Finally, P is the polarisability of the system. Here we have adopted the compressed notation $1 = (\mathbf{r}, \sigma, t)$, and 1^+ means that $t \rightarrow t + \delta$, with $\delta \rightarrow 0$ a positive number. The set of Eqs. (4.59–4.62) becomes self-consistent once we include the Dyson Eq. (4.57), linking G and Σ.

The zeroth order approximation to the vertex function assumes that quasi-electrons and quasi-holes do not interact, and vertex corrections are neglected:

$$\Gamma(1, 2, 3) = \delta(1, 2)\delta(1, 3). \tag{4.63}$$

As a result of this approximation, the polarisability is given by:

$$P(1, 2) = -iG(1, 2^+)G(2, 1). \tag{4.64}$$

Finally, for the self-energy, we obtain the approximate expression:

$$\Sigma(1, 2) = iG(1, 2)W(1^+, 2). \tag{4.65}$$

Because of the expression (4.65), the approximation that results from the first iteration of Hedin's equations as described above, is called the GW approximation.

At a practical level, one has to give G and W a specific form in order to move forward and solve Eq. (4.58) for the quasiparticle energies and wavefunctions. To do so, we follow the formulation of the GW approximation as first discussed thoroughly by Hybertsen and Louie, in their seminal paper of [25]. The authors of this work established the GW approximation for solids, showing that it provides a significant improvement to DFT, due to accounting for (dynamical) screening effects. Since then, the GW approximation has been shown to be highly accurate for predicting the quasiparticle properties of a wide class of materials, including solid state organic semiconductors [26].

4.4.2.1 Constructing the Initial Green's Function G

Returning to the GW formulation of Hybertsen and Louie, we approximate the Green's function to enter the self-energy as:

$$G(\mathbf{r}, \mathbf{r}', \omega) = \sum_{n,\mathbf{k}} \frac{\varphi_{n,\mathbf{k}}(\mathbf{r})\varphi^*_{n,\mathbf{k}}(\mathbf{r}')}{\omega - i\delta_{n,\mathbf{k}} - \varepsilon_{n,\mathbf{k}}}, \tag{4.66}$$

which follows from Eq. (4.57) for the case of a system with n eigenstates, and ignoring the many-body effects that appear in the self-energy. Here $\delta_{n,\mathbf{k}} < 0$ for $\varepsilon_{n,\mathbf{k}} > \mu$ and $\delta_{n,\mathbf{k}} > 0$ for $\varepsilon_{n,\mathbf{k}} < \mu$, with μ being the chemical potential of the system.

Within the GW approximation, the self-energy depends on the Green's function (Eq. 4.65), which in turn depends on the self-energy according to the Dyson Eq. (4.57). Therefore, even the problem of obtaining quasiparticle properties within the GW approximation is itself self-consistent, a point to which we return later. Subsequent iterations improve the accuracy of our estimate of the Green's function, however one needs to make an initial choice for the spectrum $\varepsilon_{n,\mathbf{k}}$ and wavefunctions $\varphi_{n,\mathbf{k}}$ that enter Eq. (4.66). The essential requirement is that this initial choice be reasonable, and we thus use the solutions to the DFT Kohn-Sham Eqs. (4.16). Consequently, a GW calculation always has the solution of the corresponding problem within DFT as a starting point.

4.4.2.2 Constructing the Screened Coulomb Interaction W

Regarding the calculation of the screening W, we start from the expression:

$$W(1,2) = \int d(3)\varepsilon^{-1}(1,3)v(3,2), \tag{4.67}$$

with $v = 1/|\mathbf{r} - \mathbf{r}'|$ the bare Coulomb interaction and ε the dielectric matrix of the studied material. We perform a Fourier transform of Eq. (4.67) to obtain:

$$W(\mathbf{r}, \mathbf{r}', \omega) = \sum_{\mathbf{q},\mathbf{G},\mathbf{G}'} e^{i(\mathbf{q}+\mathbf{G})\cdot\mathbf{r}} W_{\mathbf{G}\mathbf{G}'}(\mathbf{q}, \omega) e^{-i(\mathbf{q}+\mathbf{G}')\cdot\mathbf{r}'} \tag{4.68}$$

with:

$$W_{\mathbf{G}\mathbf{G}'}(\mathbf{q}, \omega) = \varepsilon^{-1}(\mathbf{q}, \omega)v(\mathbf{q}+\mathbf{G}'), \tag{4.69}$$

and reciprocal lattice vectors denoted by $\mathbf{G}$. Therefore, in order to obtain the screened Coulomb interaction, we need to calculate the dielectric matrix,which is connected to the polarisability according to the following relation:

$$\varepsilon_{\mathbf{G}\mathbf{G}'}(\mathbf{q}, \omega) = \delta_{\mathbf{G}\mathbf{G}'} - v(\mathbf{q}+\mathbf{G})P_{\mathbf{G}\mathbf{G}'}(\mathbf{q}, \omega). \tag{4.70}$$

Within linear response theory and the so-called random-phase approximation [6], the polarisability appearing in Eq. (4.70) for the dielectric matrix, corresponds to the independent particle polarisability of Eq. (4.64), which may be written as [6, 25]:

$$P_{\mathbf{G}\mathbf{G}'}(\mathbf{q}, \omega) = \sum_{n,n',\mathbf{k}} (f_{n,\mathbf{k}} - f_{n',\mathbf{k}+\mathbf{q}}) \frac{\langle n, \mathbf{k}| e^{-i(\mathbf{q}+\mathbf{G})\cdot\mathbf{r}} |n', \mathbf{k}+\mathbf{q}\rangle\langle n', \mathbf{k}+\mathbf{q}| e^{i(\mathbf{q}+\mathbf{G}')\cdot\mathbf{r}} |n, \mathbf{k}\rangle}{\varepsilon_{n,\mathbf{k}} - \varepsilon_{n',\mathbf{k}+\mathbf{q}} + \omega + i\delta} \tag{4.71}$$

Here we sum over the bands n,n' characterised by wavevectors $\mathbf{k}$,$\mathbf{k}+\mathbf{q}$. We denote the Fermi-Dirac distribution function by f. The values of P are obtained by using the solutions to the Kohn-Sham equations in the above expression. In principle one can use Eq. (4.71) to compute ε, and hence W, at finite frequencies. In practice however, performing the required Brillouin zone summations can be numerically challenging and time consuming. The approach which is taken instead, is to only compute the polarisability and the dielectric matrix from first principles at the static limit ($\omega = 0$), and then to extend the result to finite frequencies by using the so-called *plasmon pole approximation* (PPA). There exist several different versions of the PPA, as for example the one introduced by Hybertsen and Louie [25], and the one by Godby and Needs [27]. We do not wish to go into great detail regarding the PPA, and here we only present the key underlying ideas of the Hybertsen-Louie model, which are however common to other versions of PPA.

The PPA is based on the observation that the imaginary part of the dielectric function $\mathrm{Im}(\varepsilon(\mathbf{q},\omega))$ is a peaked function in ω, with the peaks corresponding to a region of finite absorption. Therefore, it is reasonable to write:

$$\mathrm{Im}^{-1}_{\mathbf{G}\mathbf{G}'}(\varepsilon(\mathbf{q},\omega)) = A_{\mathbf{G}\mathbf{G}'}(\mathbf{q})\{\delta(\omega - \tilde{\omega}_{\mathbf{G}\mathbf{G}'}(\mathbf{q})) - \delta(\omega + \tilde{\omega}_{\mathbf{G}\mathbf{G}'}(\mathbf{q}))\}, \tag{4.72}$$

with the corresponding real part:

$$\mathrm{Re}^{-1}_{\mathbf{G}\mathbf{G}'}(\varepsilon(\mathbf{q},\omega)) = 1 + \frac{\Omega^2_{\mathbf{G}\mathbf{G}'}}{\omega^2 - \tilde{\omega}^2_{\mathbf{G}\mathbf{G}'}(\mathbf{q})}. \tag{4.73}$$

Here the effective bare plasma frequency is defined as:

$$\Omega^2_{\mathbf{G}\mathbf{G}'} = \omega_p^2 \frac{(\mathbf{q}+\mathbf{G})\cdot(\mathbf{q}+\mathbf{G}')}{|\mathbf{q}+\mathbf{G}|^2} \cdot \frac{\rho(\mathbf{G}-\mathbf{G}')}{\rho(0)}, \tag{4.74}$$

with ω_p the plasma frequency and ρ the crystalline charge density.

The matrices $\tilde{\omega}_{\mathbf{G}\mathbf{G}'}$, $A_{\mathbf{G}\mathbf{G}'}$ that appear in Eqs. (4.72) and (4.73) function as parameters and have to be determined by using a set of constraints. The two necessary constraints are provided by the Kramers-Kronig relation:

$$\mathrm{Re}^{-1}_{\mathbf{G}\mathbf{G}'}(\varepsilon(\mathbf{q},\omega=0)) = \delta_{\mathbf{G}\mathbf{G}'} + \frac{2}{\pi} P \int_0^\infty d\omega \frac{1}{\omega} \mathrm{Im}^{-1}_{\mathbf{G}\mathbf{G}'}(\varepsilon(\mathbf{q},\omega)) \tag{4.75}$$

and a generalised f-sum rule:

$$\int_0^\infty d\omega\omega \mathrm{Im}^{-1}_{\mathbf{G}\mathbf{G}'}(\varepsilon(\mathbf{q},\omega)) = -\frac{\pi}{2}\omega_p^2 \frac{(\mathbf{q}+\mathbf{G})\cdot(\mathbf{q}+\mathbf{G}')}{|\mathbf{q}+\mathbf{G}|^2} \cdot \frac{\rho(\mathbf{G}-\mathbf{G}')}{\rho(0)}. \tag{4.76}$$

By using relations (4.75) and (4.76), one obtains the frequency term:

$$\tilde{\omega}^2_{\mathbf{GG}'}(\mathbf{q}) = \frac{\Omega^2_{\mathbf{GG}'}(\mathbf{q})}{\delta_{\mathbf{GG}'} - \varepsilon^{-1}_{\mathbf{GG}'}(\mathbf{q}, \omega = 0)}, \tag{4.77}$$

and the spectral function:

$$A_{\mathbf{GG}'}(\mathbf{q}) = -\frac{\pi}{2} \cdot \frac{\Omega_{\mathbf{GG}'}(\mathbf{q})}{\tilde{\omega}_{\mathbf{GG}'}(\mathbf{q})}. \tag{4.78}$$

Therefore, by computing the static dielectric matrix $\varepsilon^{-1}_{\mathbf{GG}'}(\mathbf{q}, \omega = 0)$ from first principles (based on DFT and Eq. 4.71), one can obtain the frequency terms of Eq. (4.77), and consequently the dielectric matrix at finite frequencies from Eqs. (4.72) and (4.73). Finally the screened Coulomb interaction W follows from Eq. (4.69).

4.4.2.3 Discussion of the GW Approximation

Having constructed an initial Green's function G according to Eq. (4.66), and the screened Coulomb interaction using Eq. (4.69), one can compute the self-energy operator using Eq. (4.65). Subsequently, Dyson's equation in the quasiparticle picture (Eq. 4.58) or directly for the Green's function (Eq. 4.57) is solved. Hence several properties of solids related to quasiparticles, e.g. the electronic band structure, can be obtained.

Hedin's Eqs. (4.59–4.62) along with the Dyson equation form a self-consistent problem, with the GW approximation being the first iteration towards its solution, ignoring the vertex corections that represent interactions between quasi-electrons and quasi-holes. The Green's function from this first iteration can be used to construct a new vertex function which now includes these effects, which in turn can be used to obtain an improved polarisability and self-energy and so on. However, it is worth emphasising that even the problem of obtaining quasiparticle properties within the GW approximation is itself a complex, many-body self-consistent problem. This becomes apparent from Dyson's Eq. (4.57) and the expression (4.65) for the self-energy. The self-energy depends on the Green's function, which in turn depends on the self-energy. Therefore, the approach which is usually taken is to construct an initial Green's function from the Kohn-Sham states and eigenvalues according to Eq. (4.66), and then use it to get the self-energy and an improved Green's function from the solution of the Dyson equation. This new Green's function can in turn be used to generate an improved self-energy operator, and so on. Similar to the reports in the original GW work of Hybertsen and Louie [25], for all systems studied within the GW approximation as a part of this thesis, two iterations for the Green's function are sufficient to lead to converged quasiparticle properties.

Naturally, the improved quasiparticle energies and wavefunctions obtained from the solutions of the Dyson equation can also be used to generate a more accurate screening term W, which depends on the quasiparticle spectrum through Eq. (4.71). However, for the organic systems of interest we do not pursue this self-consistency in W. The entire self-consistent procedure for obtaining quasiparticle properties within the GW approximation is depicted schematically in the flowchart of Fig. 4.1.

Finally, we would like to take a moment to discuss a physically intuitive picture regarding the superiority of GW compared to approaches such as DFT. At the static limit ($\omega = 0$), and using the expression (4.66) for G, the real part of the self-energy may be written as the sum of two terms. The first one is due to a screened exchange interaction:

$$\Sigma_{\mathrm{SEX}}(\mathbf{r}, \mathbf{r}', \omega = 0) = -\sum_{n,\mathbf{k}}^{\mathrm{occ}} \varphi_{n,\mathbf{k}}(\mathbf{r})\varphi^*_{n,\mathbf{k}}(\mathbf{r}')W(\mathbf{r}, \mathbf{r}', \omega = 0), \quad (4.79)$$

where the sum runs over all occupied states. The second is the so-called Coulomb hole term:

$$\Sigma_{\mathrm{COH}}(\mathbf{r}, \mathbf{r}', \omega = 0) = \frac{1}{2}\delta(\mathbf{r} - \mathbf{r}')[W(\mathbf{r}, \mathbf{r}', \omega = 0) - v(\mathbf{r}, \mathbf{r}')]. \quad (4.80)$$

At this static limit, the physical interpretation of the second term is clear: it represents the interaction of the quasiparticle with the induced potential due to the rearrangement of the electronic cloud caused by its presence. Capturing this effect is one of the most important differences of GW compared to DFT or Hartree-Fock. In GW the quasiparticle affects the medium within which it moves, and this in turn has an effect on its motion. In DFT and Hartree-Fock, the medium is 'rigid' and while it affects the quasiparticle properties, it is not in itself modified. One can consider the analogy of swimming for GW - the medium, i.e. water, rearranges around the swimmer, affecting their motion. On the other hand, DFT and Hartree-Fock could be described as an analogy to ice-skating, where the medium (ice), affects the motion of the skater, however its properties are not altered due to their motion.

It should be clarified that the above discussion based on the static limit of the self-energy Eqs. (4.79) and (4.80) is only used as a means of obtaining physical intuition, and the GW approximation does include finite frequency (dynamical) effects. In fact this static limit corresponds to the so-called COHSEX approximation first used by Hedin [24], upon which GW provides a significant improvement due to the inclusion of dynamical effects. This can be intuitively understood from the fact that an additional particle with a finite energy ω can potentially move fast through the medium of the screening electrons, not allowing them to rearrange in time for providing the optimal screening.

4.4.3 Excitons: The Bethe-Salpeter Equation

In order to describe exciton effects, it is not sufficient to compute the properties of independent quasi-electrons and quasi-holes. The GW approximation only describes these quasiparticles independently, as reflected by the lack of vertex corrections in Eq. (4.63). In order to describe the interaction of quasi-electrons and quasi-holes, we perform an additional iteration of Hedin's equations. By using the expression $\Sigma = iGW$ resulting from the first iteration, Eq. (4.62) for the vertex function gives:

$$\Gamma(1,2,3) = \delta(1,2)\delta(1,3) + iW(1^{+}2) \int d(67) G(1,6) G(7,2) \Gamma(6,7,3), \quad (4.81)$$

where we have resorted to the approximation $\delta\Sigma/\delta G = iW$, i.e. the derivative $iG(\delta W/\delta G)$ is neglected. Effectively, this means that the change in the screening due to an excitation is considered to be small.

In order to proceed further, one has to generalise the two-particle functions appearing in Hedin's equations, to four-particle functions that describe the creation and annihilation processes of a *pair* of quasiparticles. For example, for the screened Coulomb interaction this is achieved by writing:

$$^{4}W(1,2,3,4) = W(1,2)\delta(1,3)\delta(2,4). \quad (4.82)$$

By using the above, along with Eq. (4.81), after some algebra one obtains the generalised polarisation function from Hedin's equations [8]:

$$^{4}\bar{P} = {}^{4}P_{\mathrm{IQP}} + {}^{4}P_{\mathrm{IQP}} K \, {}^{4}\bar{P}, \quad (4.83)$$

which is the so-called *Bethe-Salpeter equation* (BSE) for excitons. This was originally introduced in 1951 by Bethe and Salpeter [28], as a way of studying the quantum mechanical bound states of a two-particle system.

Let us now discuss the different terms appearing in the BSE (4.83). The independent quasiparticle polarisability $^{4}P_{\mathrm{IQP}}$ is simply the four-particle generalisation of the polarisability resulting from the GW approximation, Eq. (4.64). The generalised polarisability $\bar{P}$ relates to the regular polarisability through the relation:

$$^{4}\bar{P}(1,2,3,4) = {}^{4}P(1,2,3,4) + \int d(5678) \, {}^{4}P(1,2,5,6)\delta(5,6)\delta(7,8)\bar{v}(5,7) \, {}^{4}\bar{P}(7,8,3,4), \quad (4.84)$$

with $\bar{v}$ being identical to the bare Coulomb interaction v for all $\mathbf{G}$ vectors, apart from $G = 0$ where $\bar{v}(\mathbf{G} = 0) = 0$. Finally, the kernel term K takes the form:

$$K(1,2,3,4) = \delta(1,2)\delta(3,4)\bar{v}(1,3) - \delta(1,3)\delta(2,4)W(1,2) = K^x + K^d \quad (4.85)$$

The first term $K^x = \delta(1,2)\delta(3,4)\bar{v}(1,3)$ appearing in the Kernel is the so-called *exchange term*, is repulsive in nature and controls aspects of the excitation spectrum such as the singlet-triplet splitting. In particular, K_x is zero for spin-triplet states, and assumes finite values for spin-singlets. This is the reason that singlet exciton states are generally more spatially extended than triplets, as discussed in detail in Chap. 6. The second term $K^d = -\delta(1,3)\delta(2,4)W(1,2)$ is referred to as the *direct term* and is the one responsible for the attractive electron-hole interaction and the formation of bound states.

At a practical level, all quantities appearing in the BSE are transformed in the basis of single-particle electron and hole wavefunctions. Along with a series of algebraic manipulations presented in [29], this allows one to write the BSE in the form of an eigenvalue problem:

$$(E_c - E_v)A^S_{vc} + \sum_{v'c'} K^{AA}_{vc,v'c'}(\Omega_S)A^S_{v'c'} = \Omega_S A^S_{vc}, \quad (4.86)$$

where the indices v, c refer to quasiparticle states in the valence and conduction bands respectively, and E_v, E_c the corresponding energies. An exciton state S has an energy Ω_S, and is written in the basis of electron-hole pairs, using the eigenvectors A^S resulting from the solution of the BSE:

$$|S\rangle = \sum_v^{\text{hole}} \sum_c^{\text{el}} A^S_{vc} |vc\rangle . \quad (4.87)$$

Essentially the coefficients A^S_{vc} quantify the contribution of the electron-hole transition $v \to c$ to the exciton state.

The terms K^{AA} represent the diagonal components of the kernel, which can generally be written in a block-diagonal form. While in principle the off-diagonal components can also play a role, these are typically ignored within the Tamm-Dancoff approximation [30], which has been discussed to only introduce small errors for a large number of different systems [29, 31–33].

The matrix elements of the direct and exchange terms of the kernel K^{AA} appearing in the BSE (4.87) are computed as:

$$K^{AA,d}_{vc,v'c'}(\Omega_S) = \int d\mathbf{r}d\mathbf{r}'\psi^*_c(\mathbf{r})\psi_{c'}(\mathbf{r})\psi_v(\mathbf{r}')\psi^*_{v'}(\mathbf{r}') \cdot \frac{i}{2\pi}\int d\omega e^{-i\omega\delta}W(\mathbf{x},\mathbf{x}',\omega)$$
$$\times\left[\frac{1}{\Omega_S - \omega - (E_{c'} - E_{v'}) + i\delta} + \frac{1}{\Omega_S + \omega - (E_c - E_v) + i\delta}\right] \quad (4.88)$$

and:

$$K^{AA,x}_{vc,v'c'}(\Omega_S) = \int d\mathbf{r}d\mathbf{r}'\psi^*_c(\mathbf{r})\psi_v(\mathbf{r})v(\mathbf{x},\mathbf{x}')\psi_{c'}(\mathbf{r}')\psi^*_{v'}(\mathbf{r}'). \tag{4.89}$$

For the calculation of the direct term (4.88), one needs the dynamically screened Coulomb interaction $W(\omega)$. In principle we could employ the plasmon pole approximation, similar to the case of a GW calculation, in order to obtain this term. However, in most semiconductor materials the differences $\Omega_S - (E_c - E_v)$ appearing in the denominator are much smaller than typical plasmon frequencies ω. This observation allows us to approximate the direct term of the kernel using the simpler form:

$$K^{AA,d}_{vc,v'c'}(\Omega_S) = -\int d\mathbf{r}d\mathbf{r}'\psi^*_c(\mathbf{r})\psi_{c'}(\mathbf{r})\psi_v(\mathbf{r}')\psi^*_{v'}(\mathbf{r}')W(\mathbf{x},\mathbf{x}',\omega=0), \tag{4.90}$$

i.e. only requiring the calculation of the static screening term, which can be obtained from first principles as outlined in Sect. 4.4.2.

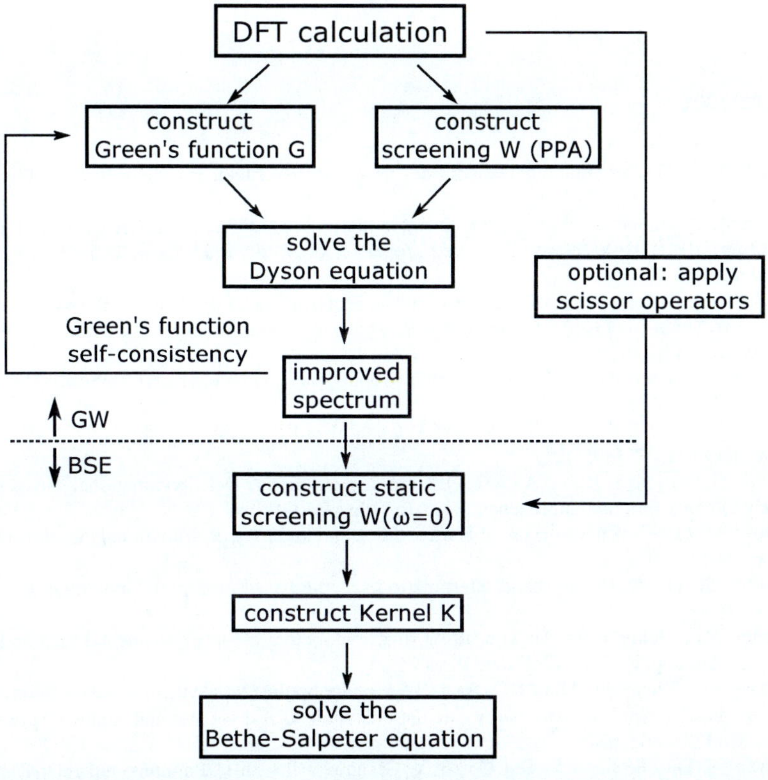

Fig. 4.1 A schematic flowchart of the GW-BSE approximation, which builds on top of a DFT calculation. The details are given in the main text

To summarise, the Bethe-Salpeter equation results from the second iteration of Hedin's equations, and describes excitons in material systems. It can be cast into the form of an eigenvalue problem (4.87), requiring input for the quasiparticle energies and a calculation of the Kernel matrix, consisting of the exchange and direct terms, given by Eqs. (4.89) and (4.90) respectively. The quasiparticle energies and wavefunctions required by this approach are most commonly obtained from a preceding GW calculation, leading to a hybrid GW-BSE approach. The steps of such a calculation are summarised in the flowchart of Fig. 4.1. However, we would like to emphasise that while it is common practice for the solution of the BSE to follow a GW calculation, this is not actually necessary, and the single-particle energies and wavefunctions can instead be provided by a DFT calculation. In fact, even in the case of performing a full GW-BSE calculation, the wavefunctions used are the Kohn-Sham wavefunctions, as it has been shown that the correction of GW might be significant when it comes to the single-particle energies, but that is not the case for the wavefunctions [25]. Hence a DFT-BSE calculation can be used in some cases, and the DFT band structure can be corrected to match those of a GW calculation, within what is referred to the application of 'scissor' operators that widen the electronic band gap.

References

1. Pauli W (1927) Zur Quantenmechanik des magnetischen Elektrons. Zeitschrift für Physik A: Hadrons und Nuclei 43:601
2. Born M, Oppenheimer JR (1927) Zur Quantentheorie der Molekeln. Annalen der Physik 84:457
3. Monserrat B (2014) On the vibrational properties of solids. PhD thesis, University of Cambridge, 2014
4. Hohenberg P, Kohn W (1964) Inhomogeneous Electron Gas. Phys Rev 136:B864
5. Kohn W, Sham LJ (1965) Self-consistent equations including exchange and correlation effects. Phys Rev 140:A1133
6. Cohen ML, Louie SG (2016) Fundamentals of condensed matter physics. Cambridge University Press, Cambridge
7. Perdew JP, Burke K, Ernzerhof M (1996) Generalized gradient approximation made simple. Phys Rev Lett 77:3865
8. Onida G, Reining L, Rubio A (2002) Electronic excitations? Density-functional versus many-body Green's-function approaches. Rev Mod Phys 74:601
9. Becke AD (1993) A new mixing of Hartree-Fock and local density-functional theories. J Chem Phys 98:1372
10. Phillips JC (1958) Energy-band interpolation scheme based on a pseudopotential. Phys Rev 112:685
11. Cohen ML, Heine V (1970) The fitting of pseudopotentials to experimental data and their subsequent application. Solid State Phys 24:37
12. Payne MC, Teter MP, Allan DC, Arias TA, Joannopoulos JD (1992) Iterative minimization techniques for ab initio total-energy calculations: molecular dynamics and conjugate gradients. Rev Mod Phys 64:1045
13. Baroni S, De Gironcoli S, Dal Corso A, Giannozzi P (2001) Phonons and related crystal properties from density-functional perturbation theory. Rev Mod Phys 73:515
14. Kunc K, Martin RM (1982) Ab initio force constants of GaAs: a new approach to calculation of phonons and dielectric properties. Phys Rev Lett 48:406

15. Feynman RP (1939) Forces in molecules. Phys Rev 56:340
16. Pariser R, Parr RG (1953) A semi-empirical theory of the electronic spectra and electronic structure of complex unsaturated molecules. I. J Chem Phys 21:466
17. Pople JA (1953) Electron interaction in unsaturated hydrocarbons. Trans Faraday Soc 49:1375
18. Kiess HG (1992) Conjugated conducting polymers. Springer-Verlag, Berlin
19. Salem 1 (1966) The molecular orbital theory of conjugated systems. Amsterdam, W.A, Benjamin, New York
20. Hubbard J (1963) Electron correlations in narrow energy bands. Proc R Soc A 276:238
21. Ohno K (1964) Some remarks on the Pariser-Parr-Pople method. Theor Chim Acta 2:219
22. Haug H, Koch SW (2004) Quantum theory of the optical and electronic properties of semiconductors, 4th edn. World Scientific
23. Hedin L, Lundqvist S (1970) Effects of electron-electron and electron-phonon interactions on the one-electron states of solids. Solid State Phys Adv Res Appl 23:1
24. Hedin L (1965) New method for calculating the one-particle green's function with application to the electron-gas problem. Phys Rev 139:796
25. Hybertsen MS, Louie SG (1986) Electron correlation in semiconductors and insulators: band gaps and quasiparticle energies. Phys Rev B 34:5390
26. Tiago ML, Northrup JE, Louie SG (2003) Ab initio calculation of the electronic and optical properties of solid pentacene. Phys Rev B 67:115212
27. Godby RW, Needs RJ (1989) Metal-insulator transition in Kohn-Sham theory and quasiparticle theory. Phys Rev Lett 62:1169
28. Salpeter EE, Bethe HA (1951) A relativistic equation for bound-state problems. Phys Rev 84:1232
29. Michael Rohlfing, Louie Steven G (2000) Electron-hole excitations and optical spectra from first principles. Phys Rev B 62:4927
30. Fetter AL (1971) Quantum theory of many particle systems. International series in pure and applied physics. McGraw-Hill, pp 538–539
31. Benedict LX, Shirley EL, Bohn RB (1998) Theory of optical absorption in diamond, Si, Ge, and GaAs. Phys Rev B 57:R9385
32. Benedict LX, Shirley EL, Bohn RB (1998) Optical absorption of insulators and the electron-hole interaction: an ab initio calculation. Phys Rev Lett 80:4514
33. Benedict LX, Shirley EL (1999) Ab initio calculation of $\varepsilon_2(\omega)$ including the electron-hole interaction: application to GaN and CaF_2. Phys Rev B 59:5441

Part II
Results

Chapter 5
First Principles Modelling of Exciton-Photon Interactions

The results presented in this chapter have been published in Ref. [1] (https://doi.org/10.1063/5.0019009), where parts of the text and figures appear. Reprinted from: Antonios M. Alvertis, Raj Pandya, Claudio Quarti, Laurent Legrand, Thierry Barisien, Bartomeu Monserrat, Andrew J. Musser, Akshay Rao, Alex W. Chin, David Beljonne. First principles modelling of exciton-polaritons in polydiacetylene chains. The Journal of Chemical Physics, 153:084103, 2020, with the permission of AIP Publishing.

The experimental work presented in this chapter was not undertaken by the author of this thesis, but by a number of collaborators who are listed as co-authors in the aforementioned publication: Raj Pandya, Laurent Legrand, Thierry Barisien, Andrew J. Musser and Akshay Rao. The measurements and their interpretation are an integral part of this combined theoretical-experimental study, and are hence presented here for completeness, along with the associated experimental methodology.

The interaction of excitons and photons is the first step to photoinduced processes and leads to a superposition state that is referred to as an exciton-polariton and describes the propagation of electromagnetic energy in matter. Here we develop a first-principles methodology for describing exciton-polaritons in organic semiconductors. This provides a more detailed picture of exciton-photon interactions, compared to the usual 'one-off' absorption picture that is considered in the weak-coupling limit. This picture becomes particularly important when exciton-photon interactions are strong, and here we explore the effect of various material parameters on their magnitude.

5.1 Introduction

Excitons are bound electron-hole pairs that appear in semiconductors, and are critical to the optoelectronic response of these materials. These quasiparticles are connected to a finite oscillating polarisation that emits an electromagnetic wave, which can act

© The Author(s), under exclusive license to Springer Nature Switzerland AG 2021

A. M. Alvertis, *On Exciton–Vibration and Exciton–Photon Interactions in Organic Semiconductors*, Springer Theses,
https://doi.org/10.1007/978-3-030-85454-6_5

on an incident wave, leading to a light-matter interplay. Once a photon is close to resonance with an exciton, light-matter coupling can become strong, and the two cannot be treated independently, as it was first realised in the pioneering works of Pekar [2, 3] and Hopfield [4] more than six decades ago. This coupling between polarisation and photons is captured by introducing the quanta of a new quasiparticle referred to as a polariton, which describes the propagation of electromagnetic energy in matter [5].

Organic semiconductors are a class of materials the light-matter interactions of which are relevant to a number of promising optoelectronic applications, making the understanding of their polariton properties interesting from a technological point of view. Their applications include light-emitting diodes (LEDs) [6], solar cells [7] and room temperature polariton lasers [8, 9]. Excitons in organic semiconductors are typically strongly bound [10, 11] Frenkel-like excitons [12–15], and their dispersion can be considered to be approximately flat [16] compared to the strong dispersion of light, as visualised in Fig. 5.1. For an incident photon close to resonance with an exciton state of an organic semiconductor, strong coupling leads to two new polariton states that are referred to as upper polariton (UP) and lower polariton (LP) respectively. As depicted in Fig. 5.1, this results in a stop-band (shaded region), where no light propagation takes place within the material. The width Δ of the stop-band depends on the magnitude of the exciton-photon interaction. The stop-band is also related to, but is distinct from, the so-called Rabi splitting that is commonly used to quantify light-matter interactions, and which is defined as the maximal vertical distance between the upper and lower polariton branches [17]. Polaritons in organic semiconductors have often been studied for molecular systems placed between highly reflecting mirrors that confine incident light [9, 18–21]. This microcavity effect leads to a standing electromagnetic wave that interacts with the material system. However, here we focus on the case of unconfined photons, and not on such microcavity polaritons.

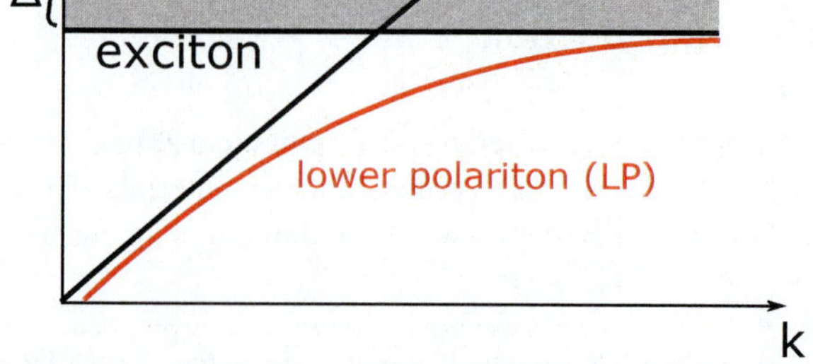

Fig. 5.1 The dispersion of a photon in a material intersects with that of an exciton, and their interaction leads to two new hybrid polariton states. The upper and lower polaritons are separated by an energy gap of width Δ (shaded region), which is referred to as the stop-band

The building block of organic semiconductors is often organic molecules, the large size and weak intermolecular van der Waals interactions of which lead to phonon modes with low frequencies, hence strongly activated at room temperature. These slow nuclear motions lead to large variations of the electronic interactions between neighbouring molecules [22, 23]. Since excitons can spatially extend over several molecules [24, 25], such low-energy phonons can have a strong influence on their properties [26]. Similarly, organic semiconductors based on conjugated polymers exhibit very low-frequency phonon modes that are strongly coupled to their excitons [27]. However, it is not only vibrations of low energies that couple strongly to excitons, as in organic systems high-frequency motions such as carbon-carbon stretches have often been discussed to dominate this interaction [28, 29]. This strong coupling of excitons to the vibrations of organic semiconductors is expected to also have an impact on exciton-photon coupling, and indeed it has been shown that vibrations have a strong effect on the dynamics of polariton states [30–34].

Therefore, understanding the properties of polaritons requires accounting not only for exciton-photon, but also exciton-vibration interactions, through a Hamiltonian operator of the form:

$$H = H_{\text{exc-ph}} + H_{\text{exc-vib}}, \tag{5.1}$$

where the operator $H_{\text{exc-vib}}$ describes the exciton-vibration interactions. For a single exciton state with energy ω_{exc}, coupled to a photon of energy ω_{ph}, the exciton-photon Hamiltonian can be written within the so-called resonance approximation [35]:

$$H_{\text{exc-ph}} = \omega_{\text{ph}}\hat{a}^{\dagger}\hat{a} + \omega_{\text{exc}}\hat{c}^{\dagger}\hat{c} - ig(\hat{a}\hat{c}^{\dagger} - \hat{a}^{\dagger}c). \tag{5.2}$$

Here $a^{\dagger}$ (a) is the creation (annihilation) operator of the photon state, and $c^{\dagger}$ (c) the creation (annihilation) operator of the exciton. The constant g quantifies the magnitude of the exciton-photon interaction and is related to the width of the stop-band Δ of Fig. 5.1 via the relationship: $g = \frac{1}{4}\sqrt{\omega_{\text{ph}}\Delta}$.

While first-principles approaches have been used in several cases to parametrize the exciton-vibration interactions encoded in $H_{\text{exc-vib}}$ [36, 37], the same cannot be said for the exciton-photon Hamiltonian of Eq. (5.2). To the best of our knowledge, a comprehensive *ab initio* method for modelling the stop-band of organic systems is thus far missing. Naturally, experiment can inform computational models on the range of values of Δ, however this does not allow for a truly predictive approach of modelling light-matter interactions in these systems. It also remains unclear how vibrational effects affect the stop-band.

Here we develop a theoretical framework for simulating exciton-photon interactions in organic systems by combining first principles calculations for excitons with the macroscopic theory of polaritons from classical electrodynamics. We thus extract the stop-band width Δ that appears in the microscopic Hamiltonian (5.2). We exemplify our approach by studying variants of the conjugated polymer polydiacetylene (PDA) (Fig. 5.2), and we show that the conjugation length of the polymer chains has a large impact on the magnitude of exciton-photon interactions. Exciton properties

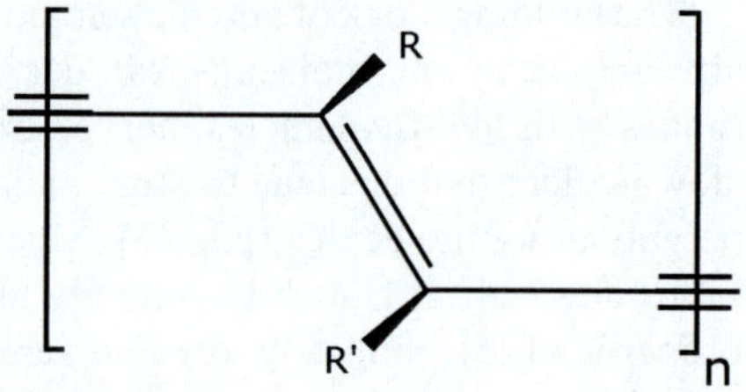

Fig. 5.2 The structure of a diacetylene monomer, which generally coincides with the unit cell of an infinite and periodic crystal of polydiacetylene

of PDA are obtained from first principles by using accurate Green's function based methods (GW-BSE) [38–40], which are known to be appropriate for the study of conjugated polymers [41], due to successfully capturing the large intrinsic correlation length of the electrons and holes that participate in exciton states. We compare the simulated results to experimental reflectivity measurements of the stop-band width, and we find the two to be in reasonable agreement. However, by only considering the effect of the main exciton peak on incident photons, the internal structure of the stop-band as it appears in reflectivity measurements is not reproduced in the calculation. By coupling GW-BSE calculations for the excitons, to non-perturbative finite differences methods for the vibrations [42, 43], we show that phonon effects have an important effect on the internal stop-band structure.

The structure of this chapter is as follows. Section 5.2 provides an overview of our theoretical approach for describing light-matter interactions. We apply our methodology to disorder-free PDA in Sect. 5.3. An overview of the studied systems is also provided in this section, alongside a discussion of their exciton properties and the experimental determination of the magnitude of light-matter interactions. We then proceed to discuss the role of static disorder in the sense of limiting the infinite conjugation length of disorder-free chains, in Sect. 5.4. The impact of dynamic disorder (vibrations) on light-matter interactions is discussed in Sect. 5.5, along with a detailed outline of the employed methodology for accounting for its effect. We summarise the entirety of the results of this chapter, and discuss potential future avenues in the conclusions of Sect. 5.6.

5.2 Theoretical Background

Consider an electromagnetic wave $\mathbf{E}$, $\mathbf{B}$ interacting with a material of dielectric $\epsilon(\omega)$. The wave equation for the electric field in frequency space reads:

$$\nabla^2\mathbf{E}(\mathbf{r},\omega) + \frac{\omega^2}{c^2}\epsilon(\omega)\mathbf{E}(\mathbf{r},\omega) = 0. \tag{5.3}$$

For the moment, and for the sake of simplicity, let us consider a bulk isotropic material, for which the electric field can be treated as a plane wave:

$$\mathbf{E}(\mathbf{r}, \omega) = E \cdot e^{i(\mathbf{k}\cdot\mathbf{r}-\omega t)} \tag{5.4}$$

By substituting Eq. (5.4) into (5.3) we obtain:

$$\left[-k^2 + \frac{\omega^2}{c^2}\epsilon(\omega)\right] E \cdot e^{i(\mathbf{k}\cdot\mathbf{r}-\omega t)} = 0. \tag{5.5}$$

The transverse eigenmodes of the system satisfy the condition $E \neq 0$, therefore:

$$c^2k^2 = \omega^2\epsilon(\omega) \tag{5.6}$$

It thus becomes obvious that if one obtains the dielectric response $\epsilon(\omega)$ of a material, the dispersion of the light field in the material, i.e. the polariton dispersion, can be obtained by solving Eq. (5.6). This method is agnostic to the level at which the electronic structure is computed, and our GW-BSE methodology provides but one possibility.

Let us now consider the so-called resonance approximation to Eq. (5.6), as a useful way of gaining physical intuition on the origins of the stop-band. However, we emphasise that we do not make use of this approximation to obtain the simulated results presented in this chapter, and it only serves as a way of developing intuition. Unless we are at resonance with an excited state $\omega = \omega_o$, the imaginary part of the dielectric response is negligible, therefore Eq. (5.6) can be approximated as:

$$c^2k^2 = \omega^2\mathrm{Re}(\epsilon(\omega)). \tag{5.7}$$

Equation (5.7) shows that there will only exist real solutions for the wavevector k once $\mathrm{Re}(\epsilon(\omega))$ is positive. While this is generally the case, $\mathrm{Re}(\epsilon(\omega))$ can become negative in the vicinity of an optical transition i with a high oscillator strength f_i. This can be seen within a Lorentz oscillator model of the dielectric response [44]:

$$\epsilon(\omega) = \epsilon_0 + \frac{Ne^2}{m}\sum_i \frac{f_i}{\omega_i^2 - \omega^2 - i\omega\gamma_i}. \tag{5.8}$$

The real part can be written as:

$$\mathrm{Re}(\epsilon(\omega)) = \epsilon_0 + \frac{Ne^2}{m}\sum_i f_i \frac{\omega_i^2 - \omega^2}{(\omega_i^2 - \omega^2)^2 + \omega^2\gamma_i^2} \tag{5.9}$$

where ϵ_0 is the background dielectric response, N the number of molecules per unit volume, m the electron mass and γ_i the dephasing rate. In the vicinity of a transition i, $\omega = \omega_i + \delta\omega$, and only keeping terms in first order of $\delta\omega$ we obtain the condition for $\mathrm{Re}(\epsilon(\omega)) < 0$ to be:

$$f_i > \left(1 + \frac{\omega_i}{2\delta\omega}\right) \cdot \frac{\gamma_i^2}{\omega_p^2}, \tag{5.10}$$

where $\omega_p^2 = \frac{Ne^2}{\epsilon_0 m}$ the plasma frequency of a material. The width of the frequency region over which condition (5.10) holds and thus $\mathrm{Re}(\epsilon(\omega)) < 0$, defines the stop-band width Δ.

Therefore, the brighter an optical transition, the more likely it is that it will lead to a forbidden region for light propagation in the material. Since the oscillator strength of a material satisfies the sum rule:

$$\sum_i f_i = Z, \tag{5.11}$$

with Z the number of electrons per molecule, we conclude that materials with fewer bright optical transitions are more likely to satisfy condition (5.10) and lead to strong light-matter interactions. This provides the motivation for studying extended one-dimensional systems like PDA, as elaborated in the following section.

Before proceeding to use the presented methodology on material systems, we would like to highlight two points regarding its range of applicability. First of all, this has been discussed in the context of a general Lorentz oscillator, and the frequencies ω_i appearing in Eq. (5.8) need not correspond to exciton states. While here we focus on exciton-polaritons, optical phonons and plasmons can also form hybrid states with light [45, 46], and the presented theoretical framework could be used to describe such phenomena. Secondly, the polariton properties we calculate correspond to the bulk of an infinite material structure. However, these states can also form at the surface of a material, only propagating along the interface between different media. These states are referred to as surface polaritons, and have a finite dispersion within the stop-band of the infinite 'bulk' polariton states that we study here [47]. Our theoretical framework does not capture such surface effects, but could be extended in that direction in the future.

5.3 Application to Disorder-Free PDA

Polydiacetylene (PDA) is a conjugated polymer with an alternating sequence of single-double-single-triple carbon-carbon bonds, as visualised in Fig. 5.2. These materials are known to give very high-quality crystals, with an almost perfect alignment of the polymer chains parallel to each other and free from conformational defects, due to the topochemical reaction they originate from [48]. Variants of PDA are traditionally named after the colour of the transmitted light through a film of the material. Here we study two PDA variants shown in Fig. 5.3, alongside their absorption and reflection spectra. We refer to the structure of Fig. 5.3a that absorbs at lower energies as 'blue' PDA [49] and the one of Fig. 5.3b that absorbs at higher energies as 'red' [50, 51]. It is important to note that the two types of PDA chains and their optical responses should in no way be associated to varying levels of disorder in their structures but are largely determined by their backbone geometries (see below) [52].

The stop-band of the two PDA variants is determined from the measured reflectivity spectra presented in Fig. 5.3, for near-normal incidence of light. Details of the experimental measurements are given in Appendix 1. If we were to disregard the absorption from the exciton over a finite region of frequencies, then every incident photon with an energy within the stop-band region would be reflected, giving a reflectance of $R = 1$ [53], since no propagation of light can take place within the forbidden region. However, the finite damping γ that appears in the dielectric function of Eq. (5.8) due to the absorption of an exciton, leads to weaker reflectance and a 'gap' within the forbidden region of the stop-band [5, 54] (as also seen in Fig. 5.7b for a model PDA system). Further gaps to the high reflectivity of the stop-band can appear due to disorder or additional bright exciton peaks. Therefore, it is not always the case that one can identify the stop-band as a region of high reflectance, and its internal structure can be rich. This is the case with the reflectivity spectra of Fig. 5.3. Nevertheless, we can still identify the width of the stop-band as the distance of two peaks in the reflectivity spectra: the one which appears on the low-energy side of the dip due to the absorption from an exciton state, and the one for which the reflectance reaches values at least as large as the maximum of the first peak. This way of determining the stop-band is similar to previous work [54], and strong light-matter coupling in squarylium [55] provides a particularly good example of using the reflectivity measurements to that end.

The presented reflectivity measurements show that the stop-band width is 2.64 and 1.16 eV for the blue and red chains respectively. We discuss the origin of this large difference between the red and blue chains with the aid of simulations below. The measured values are within the expected range, as the stop-band width has been shown to surpass values of 1–2 eV in several organic materials [54]. In particular, it has been shown that for near-normal incidence of light on a lattice of aligned dipoles, values of this order of magnitude might be obtained [54]. Given the close-to-perfect alignment of the polymer chains in PDA, as well as the large oscillator strength of their lowest bright exciton state, it is not surprising that light-matter coupling is strong.

Let us now examine some of the details of the experimental reflectivity spectra. For the blue chains, the reflectance dips around the energy of the bright singlet exciton state at 1.97 eV, and a vibronic peak is also weakly resolved at 2.15 eV. Furthermore, a small dip appears at 1.85 eV, and this has been discussed to be due to minority chain populations with different geometries, which are present within the samples [52]. Additional structure appears at energies of 2.72 and 3.06 eV. It is currently not clear what these peaks correspond to, however the side-chains that are connected to the conjugated PDA structure can host exciton states of their own, potentially leading to some of the observed features. At higher energies, the reflectance shows an increase at approximately 3.5 eV and then decreases again. This is due to exciton states of unpolymerised diacetylene monomers which are known to appear in this energy range [52]. Finally, there is a stark increase of the reflectance as photon energies approach 4.4 eV, which is then followed by a continuous drop in its value, behaviour characteristic of the stop-band edge. For the red PDA, a large dip in the reflectance appears at 2.48 eV where the bright exciton state of this structure absorbs,

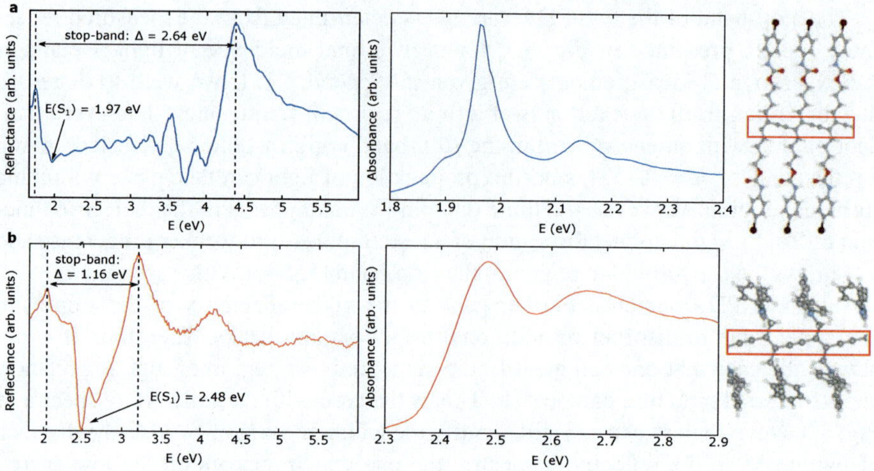

Fig. 5.3 Experimental reflection and absorption spectra of two studied variants of polydiacetylene: **a** blue chains and **b** red chains. The stop-band width of the two structures is approximately determined from the reflectivity spectra. The polymer structures are given alongside the experimental plots, with the polymer backbone highlighted in red rectangles, in order to distinguish it from the side-chains. Carbon atoms are denoted in light gray, hydrogen in white, oxygen in red, nitrogen in light blue and bromine in brown

and a vibronic peak is then present at 2.68 eV. The reflectance then increases steadily until reaching the edge of the stop-band. On the low-energy side of the main exciton peak, some additional structure is present, and similar to the case of blue PDA, we attribute this to minority chain populations.

The PDA variants of Fig. 5.3 contain a very large number of atoms in their unit cells, due to the size of the side-chains connected to the conjugated structure (highlighted with a red rectangle in Fig. 5.3). In order to understand the role of the side-chains, we resort to studying two simplified model PDA systems, where the side-chains have been substituted by hydrogen atoms, as visualised in Fig. 5.4. The small size of the model systems has the advantage of making the computational study of exciton-phonon interactions feasible, a fact that we return to later. When viewing the model PDA chains sideways as on the left of Fig. 5.4 these systems look identical. However, by looking into the direction along which the chains extend, it becomes clear that the model red structure is not planar, and there exists a finite torsional angle of 14° between subsequent diacetylene monomers. Overall, since the interatomic distances within the polymer backbone of the four studied PDA systems are taken to be identical, the differences in their optoelectronic properties are purely due to the effects of the side-groups and the (lack of) torsion between subsequent diacetylene monomers.

The effect of this torsion has been discussed in detail previously [56], and shown to lead to significant differences in the optoelectronic response of the two systems. In crystalline PDAs the side groups, through their mutual organisation, pilot the chain

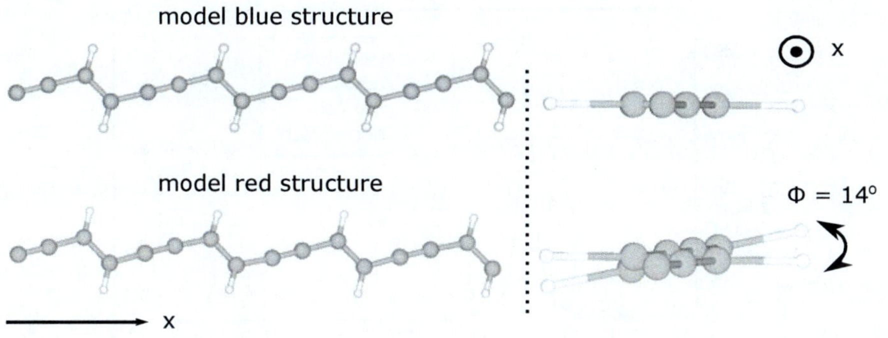

Fig. 5.4 Model PDA structures, constructed by substituting the side-chains of the real materials of Fig. 5.3 with hydrogen atoms. While the two model systems might look identical when viewed from the side (left), viewing them in the direction in which the chains lie (right) emphasises that the blue structure is planar, while a finite torsion occurs between subsequent monomers of the red structure

conformation that in turn governs the electronic properties. However, the side groups themselves do not participate to the conjugated system, justifying the substitution by hydrogen atoms. Nevertheless, this simplification leads to a renormalisation of exciton energies. This is due to the the strong dielectric mismatch introduced between the conjugated chain and the outside medium, as discussed below.

The absorption of the model PDAs and real systems as computed from GW-BSE is given in Fig. 5.5. The computational details for the electronic structure calculations are given in Appendix 2. The first thing to observe from Fig. 5.5 is that there is only a single bright transition for all the studied systems, at least in the idealised case of the infinite periodic material. Therefore, PDA provides an ideal system for satisfying condition (5.10) and observing strong light-matter interactions, since the oscillator strength is not shared between several transitions according to the sum rule of Eq. (5.11). It should however be noted that the real systems have some additional exciton states with small but non-vanishing oscillator strength at energies above 3 eV, which are associated with the side-chains.

Moreover, it is interesting to note the difference in the exciton energy between the real and the respective model structures. Since the polymer backbone is identical in both cases, it is only the side-chains that can be causing the observed difference. The side-chains of the real red PDA consist of two phenyl rings, which are highly polarisable and hence provide significant screening to the Coulomb interaction that binds electron-hole pairs together. In contrast, in the model system the field lines of this Coulomb interaction are not screened by a medium, since the side-chains have been substituted by hydrogen atoms. Consequently, the exciton is more strongly bound in the model chain, further stabilising its energy. For the real blue system, the side-groups largely consist of non-polarisable alkyl chains, hence the screening they provide is much weaker compared to the red PDA. This is reflected by the fact that the exciton energies of the model and real blue systems are close-to-identical, closely reproducing the experimental value from the absorption spectrum of Fig. 5.3a.

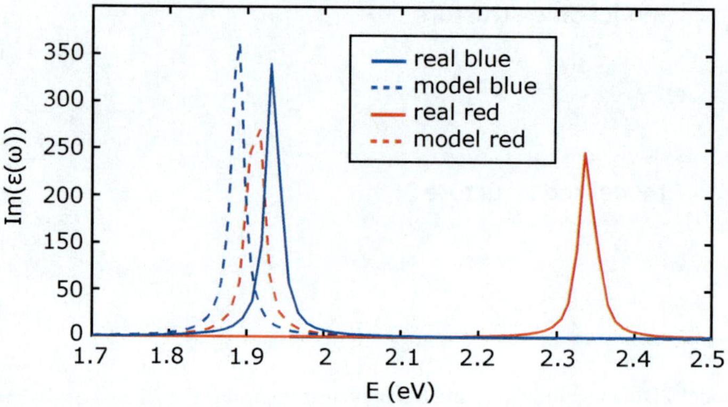

Fig. 5.5 Absorption of the real and model PDA chains

Table 5.1 Values for the first exciton energy $E(S_1)$, energy of the fundamental gap E_g and exciton binding energy $E_{binding}$ of the four PDA chains

	$E(S_1)$ (eV)	E_g (eV)	$E_{binding}$ (eV)
Real blue	1.93	3.29	1.36
Model blue	1.88	3.97	2.11
Real red	2.34	2.86	0.52
Model red	1.91	4.62	2.71

Table 5.1 summarises the values for the fundamental gaps of the chains, along with the energy of the excitons and their binding energy. The exciton binding energy of the real red and blue systems is equal to 0.52 eV and 1.36 eV respectively, while for both model chains it is greater than 2 eV, emphasising the large impact of the screening provided by the polarisable side-chains. This effect can be further emphasised by visualising the exciton wavefunctions of the real and model red chains in Fig. 5.6. The electron density is visualised in red, for a hole fixed at the position indicated by a blue sphere. As expected, the lack of side-chain screening of the Coulomb interaction between the electron and hole in the case of the model chain leads to an exciton that is spatially more confined than in the real system. As far as the relative electron-hole motion is concerned, we also reach a good agreement with the results from electroabsorption studies that provide an exciton bohr radius of several nanometers, corresponding to an exciton covering a few monomer units in the blue and red conformations [51, 57].

Having obtained $\epsilon(\omega)$ from GW-BSE electronic structure calculations of the three materials, we proceed to obtain the polariton dispersion. Due to the almost perfect alignment of PDA chains along one axis (x), we use the polariton dispersion relation [54]:

$$c^2k^2 = \omega^2\epsilon_{xx}(\omega). \tag{5.12}$$

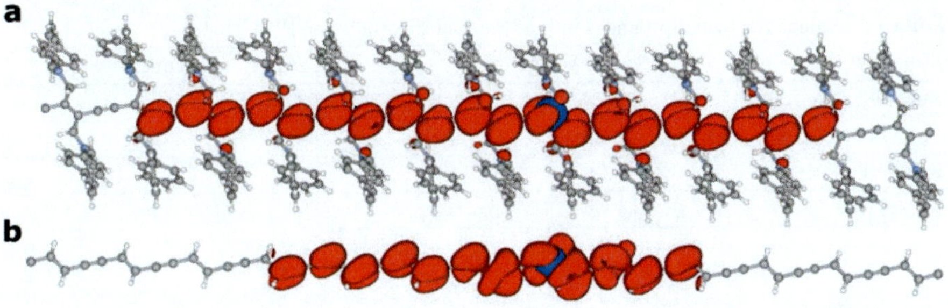

Fig. 5.6 Exciton wavefunction of **a** the real red PDA chains and **b** the model red chains. The electron density is visualised in red for a hole fixed at the position indicated by a blue sphere

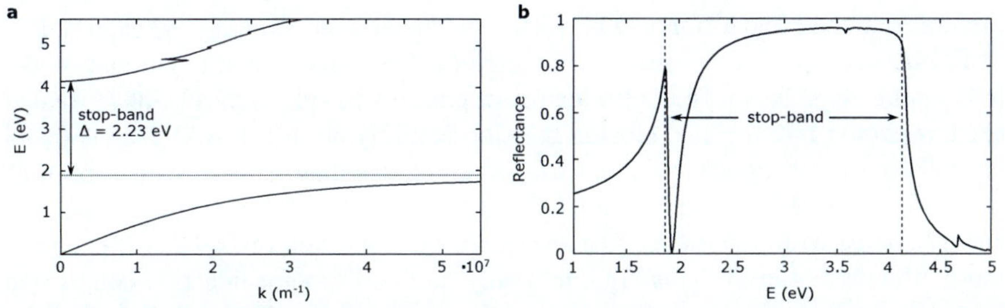

Fig. 5.7 Calculated **a** polariton dispersion and **b** reflectance of the model blue PDA chains, highlighting the stop-band of width $\Delta = 2.23\,\text{eV}$

Here $\epsilon_{xx}(\omega)$ is the only non-zero component of the dielectric tensor of the material, and in the y and z directions, the vacuum dispersion holds: $\omega^2 = c^2k^2$. We obtain the component $\epsilon_{xx}(\omega)$ of the dielectric tensor, by solving the Bethe–Salpeter equation for incident light normal to the chains, which approximately emulates the experimental conditions used to obtain the reflectivity spectra of Fig. 5.3. Figure 5.7a visualises the computed dispersion for the example case of the model blue chains, highlighting the resulting stop-band of 2.23 eV. The Δ values obtained for all studied systems are summarised in Table 5.2, and the experimental values are given for comparison. It should be emphasised that the effect of exciton dispersion on the polariton states of these systems is negligible. As outlined in Appendix 2, the bandwidth of the exciton is equal to 1.33 eV from the zone centre Γ ($|\mathbf{k}| = 0$) to the band-edge X ($|\mathbf{k}| = \frac{\pi}{\alpha} = 6.43\,\text{nm}^{-1}$) of the first Brillouin zone in the direction of the polymer chains. Within the same distance in momentum space, the energy of a photon changes by $\Delta E = \hbar c \frac{\pi}{\alpha} - 1.87 = 1268\,\text{eV}$, a change which is three orders of magnitude greater than that of the exciton energy.

We see from Table 5.2 that the model blue chains have significantly larger stop-band width than the model red system. Since the only difference between these structures is the torsional angle between subsequent diacetylene monomers (see Fig. 5.4), we conclude that the magnitude of this torsion has a strong effect on the magnitude of light-matter interactions. Indeed the finite torsion of the model red

Table 5.2 Values for the stop-band width of the four disorder-free PDA chains

Structure	Δ (eV)-theory	Δ (eV)-experiment
Real blue	2.28	2.64
Model blue	2.23	–
Real red	1.06	1.16
Model red	1.45	–

system leads to deviations from the perfect dipole alignment of model blue PDA, and which is crucial to facilitating strong coupling to photons. The computed values for the stop-band width are in reasonable agreement with the experimental values of Fig. 5.3, and the large difference between the blue and red structures is captured.

Furthermore, the comparison of the predicted Δ values for the real and model red systems suggests that the introduction of polarisable side-chains leads to weaker exciton-photon interactions, closing the stop-band by almost 0.4 eV. This is likely due to two reasons. Firstly, the dielectric screening that they provide to electron-hole pairs (Fig. 5.6) leads to a more weakly bound exciton, which has a higher transition energy compared to the model system. Hence the condition of Eq. (5.10) becomes more difficult to satisfy. This argument only holds for polarisable side-chains that provide significant screening, which is not the case for the real blue PDA. Indeed the computed values of Δ for the real and model blue systems differ by a mere 50 meV. Moreover, upon the introduction of side-groups, a few exciton states with small but non-vanishing oscillator strength appear at energies above 3 eV, hence reducing the brightness of the main exciton peak according to the sum rule of Eq. (5.11), and also making the frequency range that satisfies the condition of Eq. (5.10) narrower.

The values predicted by the calculations for the stop-band width are in reasonable agreement with experiment, as summarised in Table 5.2. We believe that the differences are largely due to our modelling of a perfect periodic structure and ignoring exciton resonances which are present in a realistic film, for example due to unpolymerised diacetylene monomers and minority chain populations at different geometries. Each of these weak transitions couples to photons and has a (narrow) stop-band of its own. The overlap of the different stop-bands due to the various weak exciton resonances with that of the main exciton peak leads to an overall wider stop-band [5]. In agreement with this line of thought, the predicted values underestimate experiment for both materials. It is however worth pointing out that while the overall stop-band is wider due to the presence of additional resonances, the individual stop-bands become narrower. This effect is discussed later and visualised in Fig. 5.12 for the main exciton peak in the presence of vibronic resonances.

For normal incidence of light, the reflectance can be calculated as:

$$R(\omega) = \frac{(n(\omega) - 1)^2}{(n(\omega) + 1)^2}, \tag{5.13}$$

where $n(\omega)$ the complex refractive index of the material. For the model blue chains this is plotted in Fig. 5.7b, where the stop-band is clearly resolved. The absorption from the exciton state results in a low-reflectance gap within the stop-band, which is in agreement with experiment. However, the computed reflectivity spectrum looks qualitatively very different from the experimental one of Fig. 5.3a. This is due to entirely ignoring the effects of disorder in the calculation. The presence of minority chain populations with different geometries, as well as unpolymerised diacetylene monomers are ignored, and the exciton states associated with these would lead to more gaps in the reflectivity. For the model chains we ignore excitons associated with the side-chains, which would also introduce additional structure within the stop-band. We will see later that dynamic disorder is another important factor towards determining the internal structure of the stop-band.

Before moving on to consider the effects of disorder, we would like to comment on the speed of energy transfer in these materials, as predicted by the results of this section. First, let us consider the case of purely excitonic motion. The quantity that we need to compute in order to quantify the speed of band-like exciton motion is the exciton group velocity:

$$v_{\mathrm{g,exc}} = \frac{1}{\hbar} \cdot \frac{\partial E_{\mathrm{exc}}}{\partial q}, \tag{5.14}$$

where E_{exc} the exciton energy and q the exciton wavenumber. Following photoexcitation, a wavepacket is generated on the exciton potential energy surface, the momentum of which is narrowly distributed around $q = 0$, since light induces close-to-vertical transitions within the Franck-Condon approximation. The centre of mass of this Γ-point exciton will then travel with the velocity of Eq. (5.14). We are interested in energy transfer in the direction of the chains, which we define to lie along the x axis, we thus approximate $v_{\mathrm{g,exc}}$ as the average along the respective dimension of reciprocal space. To this end, we calculate the exciton energy at the band-edge X, using a supercell calculation that is outlined in Appendix 2, and we obtain:

$$v_{\mathrm{g,exc}} \approx \frac{1}{\hbar} \cdot \frac{E_{\mathrm{exc}}(X) - E_{\mathrm{exc}}(\Gamma)}{q(X) - q(\Gamma)} = 3 \cdot 10^5 \text{ m/s}. \tag{5.15}$$

Now let us consider the group velocity of a polariton $v_{\mathrm{g,pol}}$ close to the exciton resonance, which can be computed from the solution of Eq. (5.12) for different energies of the incident light. The result for the model blue chains is given in Fig. 5.8a; at low frequencies, the speed of light propagating along the PDA chains is $c = c_o/n_{xx}$. However, as we approach the energy of the exciton transition, the exciton content of the lower polariton state becomes larger, reducing the speed of propagation. Figure 5.8b provides a closer view to the region around resonance, and we find that the polariton propagates with a speed of the order of 10^6 m/s, which is an order of magnitude faster than purely excitonic band-like propagation. One might observe that very close to the exciton resonance the group velocity obtains negative values, which is unphysical. This is due to the fact that in this narrow region, the material exhibits anomalous dispersion and the group velocity is generally not a useful concept in these cases [44].

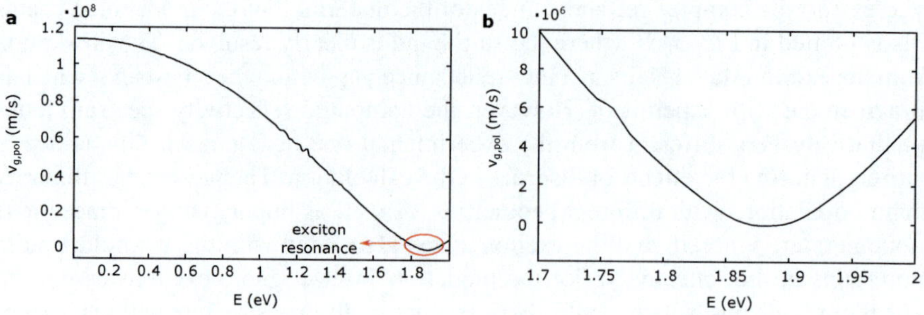

Fig. 5.8 Polariton group velocity in model blue chains, for different energies of the incident light. Panel **a** shows a large region of frequencies, whereas panel **b** provides a closer view of the region close to the resonance with the exciton

5.4 Static Disorder: The Effect of Finite Chain Length

In a realistic film of PDA, the system deviates from the perfectly periodic behaviour in more than one ways. As we saw when discussing the experimental reflectivity spectra, unpolymerised chains of diacetylene can be present, as well as minority populations of chains at different conformations. Imperfections in the chain polymerisation can lead to breaks in the conjugation of the system, or even to the termination of the polymer structure after a finite number of monomers. Here we consider such effects limiting the chain conjugation, and we ignore atomic defects and vacancies, due to the pure character of the samples as discussed previously.

In order to understand the impact of finite conjugation lengths on the exciton and polariton properties, we study five different finite systems consisting of one to five diacetylene monomers, which we terminate on both sides with hydrogen atoms. For simplicity, we choose to terminate the chains after the occurrence of a triple bond, so that the carbon atoms at the boundaries only need to bond with a single hydrogen. As an example, Fig. 5.9 visualises a trimer of diacetylene.

We perform GW-BSE calculations on these oligomers, introducing a vacuum of 20 Å between their periodic images. We thus obtain $\epsilon(\omega)$ for each of these, and similar to the previous section, compute the energy of the first singlet exciton $E(S_1)$

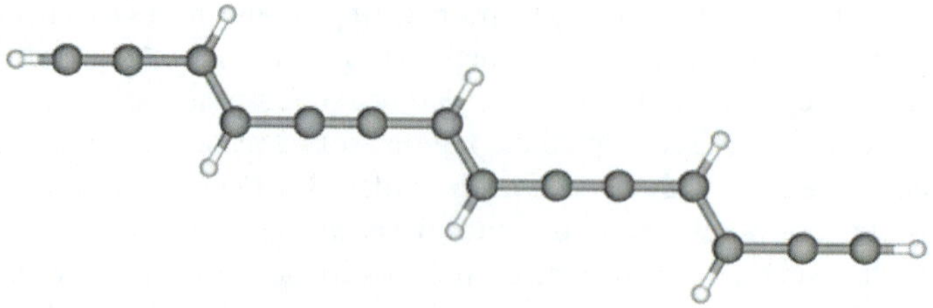

Fig. 5.9 A hydrogen-terminated trimer of diacetylene

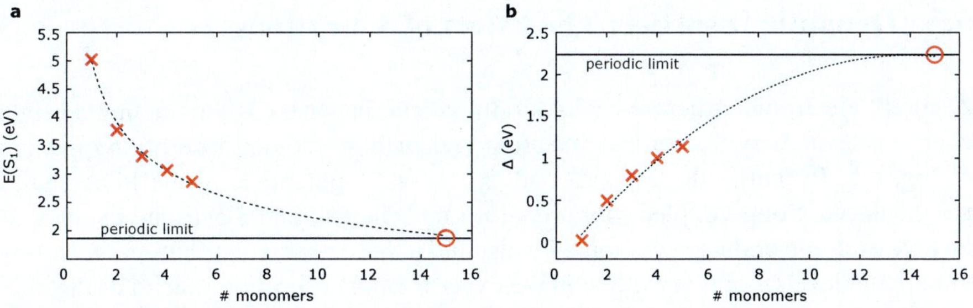

Fig. 5.10 **a** Exciton energy and **b** stop-band width for chains consisting of a different number of diacetylene monomers N. Since the constructed oligomers are planar, the periodic limit of the model blue chains is approached for a larger number of monomers

and the stop-band width Δ, which we visualise in Fig. 5.10a, b respectively. We have chosen to construct the oligomer chains so that they are planar, and hence by increasing the number of monomers, we approach the periodic limit of the model blue PDA chains of the previous section, which is also noted on Fig. 5.10.

First of all, it is clear from Fig. 5.10a that as we add more monomers to the chain, the exciton energy is stabilised and converges to the periodic limit. Due to the large computational cost, we were unable to add the required number N of diacetylene monomers in order to reach the periodic limit, however by performing model fits to the first five datapoints of Fig. 5.10a, b (dashed line), we estimate that the periodic limit is obtained for N =13–15 monomers, in very close agreement with a previous experimental study [58]. Furthermore, we find that the stop-band width increases with a larger number of monomers in Fig. 5.10b, and the periodic limit of an infinite chain provides the maximal possible magnitude for light-matter interactions. This can be rationalised from the fact that the exciton frequency appearing in the oscillator strength condition (5.10) decreases with increasing number of diacetylene monomers.

We therefore conclude that the presence of static disorder, in this case in the form of finite chain lengths, reduces the magnitude of light-matter interactions. However, the large measured stop-band width of $\Delta = 2.64\,\text{eV}$ for the blue chains suggests that conjugation-limiting effects are not important in PDA. Naturally, breaks in the conjugation and imperfect polymerisation do occur, however the results suggest that the vast majority of chains extend beyond 15 diacetylene monomers, for which number the periodic limit is reached, and the maximal light-matter coupling is obtained. Here we do not compute the effect of static disorder on the reflectance of PDA, as this would require explicit inclusion of e.g. unpolymerised diacetylene monomers within a supercell of PDA, making the cost of these calculations prohibitive. We now proceed to discuss the effects of dynamic disorder on light-matter coupling.

5.5 Dynamic Disorder: The Effect of Vibrations

A purely electronic structure calculation predicts that there is only a single bright exciton state in PDA, the vertical transition peak in the computed absorption spectrum of Fig. 5.5. In reality the ground- and excited-state potentials of the PDA chains are displaced along coupled vibrational normal modes, and electronic excitations involve multiple ending vibrational levels (and a thermalised distribution of starting vibrational levels). This is evident in the experimental absorption spectra of Fig. 5.3, where one can resolve vibrationally dressed 0–1 and 0–2 phonon side bands on top of the purely 0–0 electronic transition. Intuitively, it is obvious that these phonon side bands in the structure of $\mathrm{Im}(\epsilon(\omega))$, will give rise to more structure in $\mathrm{Re}(\epsilon(\omega))$ compared to the case of a single exciton peak, according to the Kramers–Kronig relation:

$$\mathrm{Re}(\epsilon(\omega)) = P\int_{0}^{+\infty}\frac{d\nu}{\pi}\mathrm{Im}(\epsilon(\nu))\frac{2\nu}{\nu^2-\omega^2} \tag{5.16}$$

This effect reduces the width of the region in which $\mathrm{Re}(\epsilon(\omega)) < 0$ following an excitonic transition, as we show in Fig. 5.11b. We call the width of this region Δ_{vib}, and the fact that it is smaller than Δ means that a gap in the high reflectivity of the stop-band is introduced (see Fig. 5.11c). We therefore proceed to study the effects of exciton-phonon coupling, as a means of understanding the internal structure of the stop-band. To this end, we employ the linear-vibronic Hamiltonian [59]:

$$H = \mathrm{E}(\mathrm{S}_1)\cdot I + \sum_k \hbar\omega_k\lambda_k(b_k^\dagger + b_k) + \sum_k \hbar\omega_k b_k^\dagger b_k, \tag{5.17}$$

where I is the identity, k runs over the phonon modes of the system, which have frequencies of ω_k. The energy of the first singlet exciton refers to the infinite periodic limit. The Huang–Rhys factors are calculated as:

$$\lambda_k^2 = \left(\frac{\mathrm{V}_{\mathrm{ep},k}}{\hbar\omega_k}\right)^2, \tag{5.18}$$

where we approximate the first derivative $\mathrm{V}_{\mathrm{ep},k}$ of the exciton energy along mode k using the finite difference formula:

$$\mathrm{V}_{\mathrm{ep},k} = \frac{\mathrm{E}(\mathrm{S}_1)_{+\delta u(k)} - \mathrm{E}(\mathrm{S}_1)_{-\delta u(k)}}{2\delta u(k)}, \tag{5.19}$$

with $\delta u(k)$ being a small dimensionless displacement along mode k. At a practical level, the calculation of $\mathrm{V}_{\mathrm{ep},k}$ requires two GW-BSE calculations at the displaced configurations of each mode, and constitutes a very computationally intensive process. We were therefore only able to obtain a full parametrisation of the Hamiltonian

(5.17) for the two model chains, which only have a small number of phonon modes due to the small size of their unit cell (18 and 36 modes for the model blue and red chains respectively). The existence of the side-chains for the real PDAs, for which we previously calculated the exciton properties, makes the task of computing exciton-phonon interactions a very formidable one, and we thus limit the following analysis to the model chains.

When computing the Huang–Rhys factors of PDA, one needs to account for the fact that the exciton can travel a finite distance of N monomers coherently, before being scattered by a phonon. Therefore if $\lambda_{k,\mathrm{unit}}$ is the value obtained within a unit cell exciton calculation where the phonon mode k has been displaced, then the actual value of the Huang–Rhys parameter is [60]:

$$\lambda_k^2 = \frac{\lambda_{k,\mathrm{unit}}^2}{N}. \tag{5.20}$$

We estimate the exciton coherence length N as:

$$N = \frac{v_{\mathrm{g,exc}} \cdot \tau_{\mathrm{scatter}}}{L_{\mathrm{DA}}}, \tag{5.21}$$

where $v_{\mathrm{g,exc}}$ the exciton group velocity from Sect. 5.3, τ_{scatter} the timescale for the exciton to be scattered by a phonon, and L_{DA} the length of a single diacetylene monomer. Since we find that the excited state surface of the double-bond stretch is the steepest, i.e. has the largest value of $\mathrm{V}_{\mathrm{ep},k}$, we approximate τ_{scatter} by the period of 23 fs of this mode, giving us an exciton coherence length of $N = 15$ monomers for both model chains, in good agreement with previous estimates on similar systems [29].

We write the ith eigenstate of the Hamiltonian of Eq. (5.17) as:

$$|\Psi^{(i)}\rangle = \sum_{k=1}^{M} \sum_{\tilde{v}}^{N_q} c_{\tilde{v}_k}^{(i)} |1, \tilde{v}_k\rangle \tag{5.22}$$

where there are M vibrational modes in total. The basis state $|1, \tilde{v}_k\rangle$ is the short-hand notation for

$$|1, \tilde{v}_k\rangle = |1, 0, 0, ..., \tilde{v}_k, 0, ..., 0\rangle, \tag{5.23}$$

i.e. the state where the exciton is populated and the first index set to one, and where among all vibrational modes, only mode k is populated by $\tilde{v}_k$ quanta. We thus only allow a single mode to be excited at one time, as a means of simplifying the problem. In principle, every phonon mode can admit an infinite number of quanta, however in practice we need to truncate the associated Hilbert space to only allow for up to N_q quanta in each mode. We find that for the model PDA chains $N_q = 3$ converges the results for the dielectric response.

Similar to previous works [59], we include the effect of vibrations in the absorption spectrum by taking:

$$\mathrm{Im}(\epsilon(\omega)) = (c_0^{(1)})^2 \cdot \mathrm{Im}(\epsilon(\omega))_{0-0} + \left| \sum_k \sum_{\tilde{v}} c_{\tilde{v}_k}^{(1)} \frac{\lambda_k^{2\tilde{v}_k} e^{-\lambda_k^2}}{\tilde{v}_k!} \right|^2 W(\omega - \omega_{S_1} - \omega_k), \tag{5.24}$$

where $W(\omega - \omega_{S_1} - \omega_k)$ a Lorentzian centered around $\omega_{S_1} + \omega_k$, for which we use the same width as for the peaks of the purely electronic part $\mathrm{Im}(\epsilon(\omega))_{0-0}$ of the dielectric response, as obtained from GW-BSE calculations detailed in Appendix 2. Moreover, in Appendix 2, we show for the example case of the model blue chains, that the result for the width Δ_{vib} is largely independent of the peak width γ, for values of γ between 5 and 50 meV. The coefficients $c^{(1)}$ are obtained by diagonalizing the Hamiltonian (5.17), with $c_0^{(1)}$ denoting the contribution from the purely excitonic basis state $|1, 0, 0, ..., 0\rangle$. Here we resort to approximating the absorbance $\alpha(\omega)$ with $\mathrm{Im}(\epsilon(\omega))$. This is a reasonable approximation for determining Δ_{vib}, since this quantity depends on the intensity ratio of the 0–0, 0–1, 0–2 peaks etc. The absorbance $\alpha(\omega)$ and imaginary dielectric response $\mathrm{Im}(\epsilon(\omega))$ are proportional to each other, which suggests that the error we introduce is small.

For the example case of the model blue chains, we visualise the result of including phonon effects on $\mathrm{Im}(\epsilon(\omega))$ in Fig. 5.11a, while Fig. 5.11b shows the corresponding $\mathrm{Re}(\epsilon(\omega))$ as obtained by using the Kramers–Kronig relation (5.16). For $\mathrm{Re}(\epsilon(\omega))$ we also draw a comparison to the case without phonons. The peaks that feature prominently in the absorption spectrum are due to the double- and triple-bond carbon stretching motions, with energies of 0.18 eV and 0.26 eV respectively. Indeed in the experimental absorption spectrum of the blue PDA in Fig. 5.3a, two vibronic peaks feature prominently at energies of 0.19 and 0.25 eV above the main exciton peak, however their relative intensity is opposite to the computed one. Intuitively, one can understand why the double-bond and triple-bond vibrations are most strongly coupled to the exciton from the fact that they transiently lead to structures with a smaller bond-length alternation, i.e. deviating from the Peierls-distorted structure that gives conjugated polymers such as PDA their semiconducting character, and bring the material closer to the metallic limit [61]. It is hence not surprising that these motions have a large effect on the optical properties of the studied PDAs. This also suggests that the coupling to these motions is extremely sensitive to the precise bond-length alternation of the PDA backbone, potentially accounting for the discrepancy between the theoretical and experimental absorption spectra when it comes to their relative importance. For the computed exciton coherence length of $N = 15$, we obtain Δ_{vib} values of 137 and 123 meV for the model blue and the model red chains respectively. We compute the associated reflectivity spectra in Fig. 5.11c, where we find that vibronic peaks introduce 'gaps' within the high reflectance region of the stop-band.

Therefore, accounting for vibrational effects allows one to predict part of the internal structure of the stop-band of an exciton that is strongly coupled to a photon. However, the precise form of the reflectivity, depends very sensitively on the exciton coherence length N, as shown in Fig. 5.12, where we use the width Δ_{vib} as a proxy

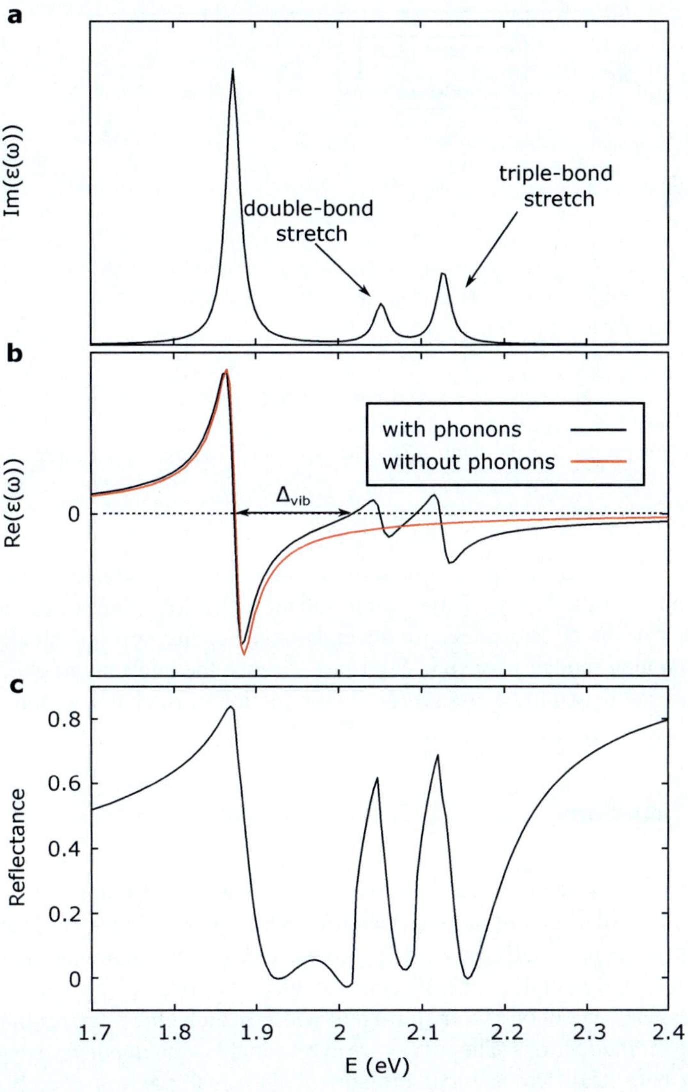

Fig. 5.11 **a** Imaginary and **b** real part of the dielectric response of the model blue chains, once the effect of phonons is accounted for. The main extra contributions to the structure of $\epsilon(\omega)$, compared to the disorder-free case, arise due to the effect of the carbon-carbon double- and triple-bond stretch. Comparing the structure of $\mathrm{Re}(\epsilon(\omega))$ for the cases with and without phonons, we find that dynamic disorder leads to 'breaks' within the region of $\mathrm{Re}(\epsilon(\omega)) < 0$. This leads to a richer structure for the reflectance in panel **c**, compared to the disorder-free case of Fig. 5.7

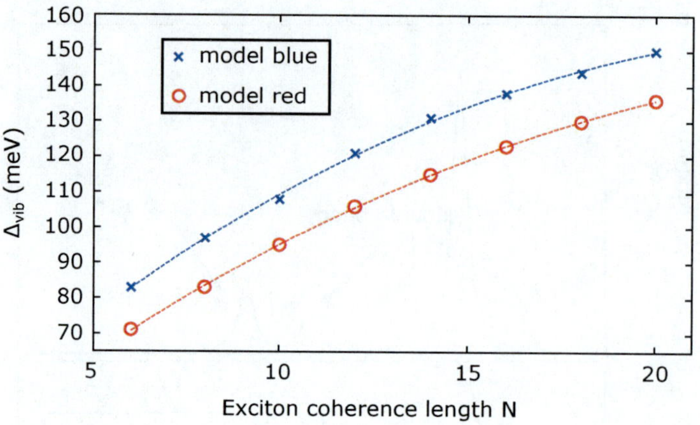

Fig. 5.12 The dependence of the width Δ_{vib} of the two model PDA chains, on the exciton coherence length

for the size of phonon-induced gaps in the reflectivity. This dependence of Δ_{vib} on N enters through the Huang–Rhys factors of the extended system, see Eq. (5.20). In turn, the exciton coherence length depends on the exciton group velocity and the period of phonon modes, as seen in Eq. (5.21). Hence the interplay of phonons with excitons is established to be crucial for the interaction of the latter with light.

5.6 Conclusions

In this work, we have presented a theoretical framework for obtaining the properties of exciton-polaritons in organic materials. Our approach requires input from a calculation of the material's dielectric response, and is agnostic to the methodology used to obtain it. By employing GW-BSE calculations for the electronic structure of polydiacetylenes, we obtain their stop-band width, which characterises the strength of light-matter interactions. The results are in reasonable agreement with experiment, and we attribute the differences to our treatment of the materials as perfectly periodic, neglecting effects of disorder and additional excitons that weakly couple to light. We compute the speed of propagation of a polariton close to the resonance of light with an exciton, showing that this is an order of magnitude faster than purely excitonic motion. We find the conjugation length of polymer chains to be a critical parameter towards determining the magnitude of light-matter interactions, highlighting one manifestation of how static disorder can lead to weaker exciton-photon coupling. This effect is not important for the polydiacetylenes studied here, due to the purity of these materials. It could however be of relevance for different systems. Furthermore, we discuss the important role of dynamic disorder and show that vibrational effects introduce gaps within the frequency range of large reflectance that characterises polaritons. We also discuss the screening of the Coulomb interaction provided by the side-chains of polydiacetylene as another potentially important contribution towards determining the magnitude of light-matter interactions. Thus all of the aforementioned factors appearing in realistic materials are important towards an accurate description of exciton-photon coupling. In principle, one needs to account for the above factors at the same time and self-consistently, something which is computationally challenging and will be the subject of a future work.

Appendix 1: Experimental Reflectivity Measurements

While the experimental work presented in this chapter was not undertaken by the author of this thesis, the relevant details are given here for completeness.

Microscopic transmission and reflectance measurements were performed using a Zeiss AxioScope optical microscope in Köhler illumination equipped with a 100× objective (Zeiss EC Epiplan-APOCHROMAT 0.95 HD DIC) coupled to a spectrometer (Avantes HS2048) via an optical fibre (Thorlabs, FC-UV50-2-SR). Five spectra were collected for each sample using an integration time of 10 and 20 ms for reflection and transmission measurements, respectively. The reflectance and transmittance were calculated using a silver mirror (ThorLabs, PF10-03-P01) and the glass substrate as references respectively.

Appendix 2: Computational Details

2.1 Electronic Structure Calculations

We perform DFT calculations using the Quantum Espresso software package [62], with the PBE functional together with a 60 Ry plane wave cutoff energy. For the model blue and model red chains, we employ $8 \times 4 \times 2$ and $4 \times 4 \times 2$ $\mathbf{k}$-point grids respectively. For the real red chains, we use $4 \times 4 \times 2$, and for the real blue system $4 \times 2 \times 4$. For the model PDAs, we introduce a vacuum of 10 Å between neighbouring polymer chains, in order to minimise inter-chain interactions.

We perform energy self-consistent GW calculations (i.e. GW_0) using the YAMBO code [63]. For the real chains, we find that including 400 Kohn–Sham states, 300 bands for the calculation of the polarisation function, and a 7 Ry cutoff for the dielectric matrix, leads to converged values for the quasiparticle band gaps. For the model blue (red) chains, we use 40 (80) Kohn–Sham states, 40 (80) bands for the calculation of the polarisation function, and a 3 Ry cutoff for the dielectric matrix.

We solve the Bethe–Salpeter equation for the model blue (red) PDA using 4 (8) occupied and 4 (8) unoccupied bands, converging the position of the first exciton peak. For the exchange term in the Bethe–Salpeter kernel, the cutoff is set to 60 Ry. For the real chains, we use 10 occupied and 10 unoccupied bands, and set the exchange to 40 Ry. In order to emulate the experimental conditions, the incident electric field is chosen to lie in the direction of the polymer chains. The imaginary part of $\epsilon(\omega)$ is obtained by fitting a Lorentzian of 10 meV width to the computed excitonic eigenvalues of the BSE Hamiltonian. The broadening of the absorption spectrum due to vibrational effects as presented in Fig. 5.11 is then added on top of that, along with the additional peaks.

Due to the one-dimensional character of the model PDAs, divergences can occur at small values of q (i.e. at long distances) in the Coulomb term that appears in the many-body calculations. In order to avoid this problem, we use the random integration method, as implemented in YAMBO. We converge the calculations using 10^5 random q-points in the first Brillouin zone, and by setting the cutoff for the real space components of the Coulomb interaction to 20 Ry.

For the model PDAs, we estimate the dispersion of the exciton band along $\Gamma - X$ as follows. We generate a supercell of size $2 \times 1 \times 1$, thus halving the size of the first Brillouin zone. As a result, the zone-boundary X point folds onto Γ, which appears as an extra exciton state once the BSE Hamiltonian of the $2 \times 1 \times 1$ system is diagonalised. The distance of this new band from the exciton value at Γ gives the width of the exciton band, which we find to be equal to 1.33 eV.

The aforementioned electronic structure calculations are performed on the PDA structures as obtained from X-ray diffraction (for the real systems). For the model systems, calculations are performed on the structures as obtained from X-ray diffraction and following the substitution of the side-chains with hydrogen atoms. Due to the extreme sensitivity of the computed exciton properties on the bond-length alternation (BLA) of the polymer chain, we set the BLA of the blue chains to be exactly identical to that of the red chains, in order to focus on the effects of the relative torsion between diacetylene monomers and on the screening effect provided by the side-chains. A geometry optimisation of these structures is not carried out, as the BLA of these one-dimensional carbon chains is known to be extremely sensitive to the fraction of exact exchange that the employed DFT functional includes [64]. It has been shown that while hybrid functionals perform better than GGA [64], it is necessary to use very accurate methods such as Quantum Monte Carlo in order to obtain geometries with realistic BLAs, which is beyond the scope of this work. Since the exciton energy is extremely sensitive to the precise value of the BLA, we limit ourselves to obtaining exciton energies for unoptimised structures, and only optimise the geometry of the studied structures prior to phonon calculations.

2.2 Phonon Calculations

We calculate the lattice dynamics of the different PDA variants in the harmonic approximation using finite differences [42, 43]. We calculate the dynamical matrix on a $1 \times 1 \times 1$ coarse **q**-point grid. Therefore, we only account for the effect of Γ-

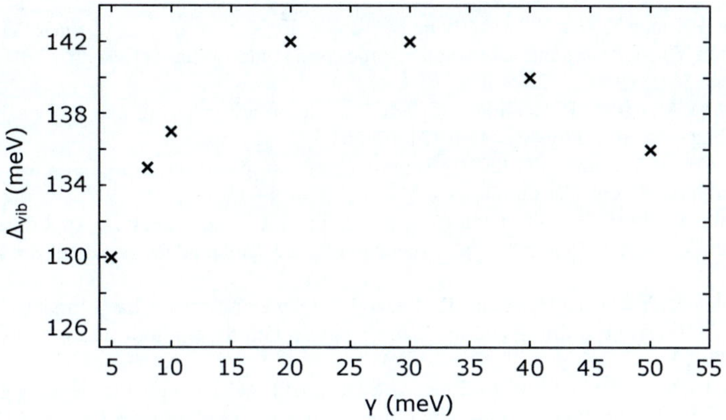

Fig. 5.13 The dependence of the width Δ_{vib} of the first region with $\text{Re}(\epsilon(\omega))$ on the width γ of the exciton and vibronic peaks

point phonons, which provides a good approximation, since these are most strongly coupled to the zero-momentum exciton states that result from photoexcitation. Prior to phonon calculations, the geometry of the different PDAs is optimised using the PBE functional.

2.3 Dependence of Polariton Splitting on Peak Width

Figure 5.13 visualises the value of Δ_{vib} of the model blue PDA, within a range of values for the width of the exciton and vibronic peaks, ranging from 5 to 50 meV. We find that Δ_{vib} does not vary by more than 10 meV within this range of widths. Here an exciton coherence length $N = 15$ has been used.

References

1. Alvertis AM, Pandya R, Quarti C, Legrand L, Barisien T, Monserrat B, Musser AJ, Rao A, Chin AW, Beljonne D (2020) First principles modeling of exciton-polaritons in polydiacetylene chains. J Chem Phys 153:084103
2. Pekar S (1958) The theory of electromagnetic waves in a crystal in which excitons are produced. Soviet J Exp Theor Phys 6:785
3. Pekar SI (1960) Identification of excitons with light waves in a crystal. Soviet J Exp Theor Phys 11:1286
4. Hopfield JJ (1958) Theory of the contribution of excitons to the complex dielectric constant of crystals. Phys Rev 112:1555
5. Kalt H, Klingshirn CF (2019) Semiconductor optics 1, 5th edn, Chap 8. Springer
6. Reineke S, Lindner F, Schwartz G, Seidler N, Walzer K, Lüssem B, Leo K (2009) White organic light-emitting diodes with fluorescent tube efficiency. Nature 459:234

7. Meng L, Zhang Y, Wan X, Li C, Zhang X, Wang Y, Ke X, Xiao Z, Ding L, Xia R, Yip H-L, Cao Y, Chen Y (2018) Organic and solution-processed tandem solar cells with 17.3% efficiency. Science 361:1094
8. Strashko A, Kirton P, Keeling J (2018) Organic polariton lasing and the weak to strong coupling crossover. Phys Rev Lett 121:193601
9. Kéna-Cohen S, Forrest SR (2010) Room-temperature polariton lasing in an organic single-crystal microcavity. Nat Photonics 4:371
10. Knupfer M (2003) Exciton binding energies in organic semiconductors. Appl Phys A 77:623
11. Köhler A, Bässler H (2009) Triplet states in organic semiconductors. Mater Sci Eng R Rep 66:71
12. Beljonne D, Yamagata H, Brédas JL, Spano FC, Olivier Y (2013) Charge-transfer excitations steer the Davydov splitting and mediate singlet exciton fission in pentacene. Phys Rev Lett 110:226402
13. Berkelbach TC, Hybertsen MS, Reichman DR (2013) Microscopic theory of singlet exciton fission. II. Application to pentacene dimers and the role of superexchange. J Chem Phys 138:114103
14. Yost SR, Lee J, Wilson MWB, Wu T, McMahon DP, Parkhurst RR, Thompson NJ, Congreve DN, Rao A, Johnson K, Sfeir MY, Bawendi MG, Swager TM, Friend RH, Baldo M, Van Voorhis T (2014) A transferable model for singlet-fission kinetics. Nat Chem 6:492
15. Coto PB, Sharifzadeh S, Neaton JB, Thoss M (2015) Low-lying electronic excited states of pentacene oligomers: a comparative electronic structure study in the context of singlet fission. J Chem Theory Comput 11:147
16. Litinskaya M, Reineker P, Agranovich VM (2006) Exciton-polaritons in organic microcavities. J Lumin 119–120:277
17. Lidzey DG, Bradley DDC, Skolnick MS, Virgili T, Walker S, Whittaker DM (1998) Strong exciton-photon coupling in an organic semiconductor microcavity. Nature 395:53
18. Polak D, Jayaprakash R, Lyons TP, Martínez-Martínez L, Leventis A, Fallon KJ, Coulthard H, Bossanyi DG, Georgiou K, Petty AJ, Anthony J, Bronstein H, Yuen-Zhou J, Tartakovskii AI, Clark J, Musser AJ (2020) Manipulating molecules with strong coupling: harvesting triplet excitons in organic exciton microcavities. Chem Sci 11:343
19. Agranovich VM, Litinskaia M, Lidzey DG (2003) Cavity polaritons in microcavities containing disordered organic semiconductors. Phys Rev B 67:085311
20. Canaguier-Durand A, Genet C, Lambrecht A, Ebbesen TW, Reynaud S (2015) Non-Markovian polariton dynamics in organic strong coupling. Eur Phys J D 69:24
21. Ebbesen TW (2016) Hybrid light-matter states in a molecular and material science perspective. Acc Chem Res 49:2403
22. Troisi A, Orlandi G, Anthony JE (2005) Electronic interactions and thermal disorder in molecular crystals containing cofacial pentacene units. Chem Mater 17:5024
23. Troisi A, Orlandi G (2006) Dynamics of the intermolecular transfer integral in crystalline organic semiconductors. J Phys Chem A 110:4065
24. Cudazzo P, Sottile F, Rubio A, Gatti M (2015) Exciton dispersion in molecular solids. J Phys Condens Matter 27
25. Refaely-Abramson S, Da Jornada FH, Louie SG, Neaton JB (2017) Origins of singlet fission in solid pentacene from an ab initio Green's function approach. Phys Rev Lett 119:267401
26. Alvertis AM, Pandya R, Muscarella LA, Sawhney N, Nguyen M, Ehrler B, Rao A, Friend RH, Chin AW, Monserrat B (2020) Impact of exciton delocalization on exciton-vibration interactions in organic semiconductors. Phys Rev B 102:081122(R)
27. Mahrt RF, Bässler H (1991) Light and heavy excitonic polarons in conjugated polymers. Synth Met 45:107
28. Spano FC (2010) The spectral signatures of Frenkel polarons in H- and J-aggregates. Acc Chem Res 43:429
29. Yamagata H, Spano FC (2011) Vibronic coupling in quantum wires: applications to polydiacetylene. J Chem Phys 135:054906

30. Somaschi N, Mouchliadis L, Coles D, Perakis IE, Lidzey DG, Lagoudakis PG, Savvidis PG (2011) Ultrafast polariton population build-up mediated by molecular phonons in organic microcavities. Appl Phys Lett 99:143303
31. Wu N, Feist J, Garcia-Vidal FJ (2016) When polarons meet polaritons: exciton-vibration interactions in organic molecules strongly coupled to confined light fields. Phys Rev B 94:195409
32. Coles DM, Michetti P, Clark C, Tsoi WC, Adawi AM, Kim JS, Lidzey DG (2011) Vibrationally assisted polariton-relaxation processes in strongly coupled organic-semiconductor microcavities. Adv Funct Mater 21:3691
33. Del Pino J, Schröder FAYN, Chin AW, Feist J, Garcia-Vidal FJ (2018) Tensor network simulation of polaron-polaritons in organic microcavities. Phys Rev B 98:165416
34. Del Pino J, Schröder FAYN, Chin AW, Feist J, Garcia-Vidal FJ (2018) Tensor network simulation of non-Markovian dynamics in organic polaritons. Phys Rev Lett 121:227401
35. Haug H, Koch SW (2004) Quantum theory of the optical and electronic properties of semiconductors, 4th edn. World Scientific
36. Schröder FAYN, Turban DHP, Musser AJ, Hine NDM, Chin AW (2019) Tensor network simulation of multi-environmental open quantum dynamics via machine learning and entanglement renormalisation. Nat Commun 10:1062
37. Alvertis AM, Schröder FAYN, Chin AW (2019) Non-equilibrium relaxation of hot states in organic semiconductors: impact of mode-selective excitation on charge transfer. J Chem Phys 151:084104
38. Lars H (1965) New method for calculating the one-particle Green's function with application to the electron-gas problem. Phys Rev 139:796
39. Rohlfing M, Louie SG (1998) Electron-hole excitations in semiconductors and insulators. Phys Rev Lett 81:2312
40. Rohlfing M, Louie SG (2000) Electron-hole excitations and optical spectra from first principles. Phys Rev B 62:4927
41. Rohlfing M, Louie SG (1999) Optical excitations in conjugated polymers. Phys Rev Lett 82:1959
42. Kunc K, Martin RM (1982) Ab initio force constants of GaAs: a new approach to calculation of phonons and dielectric properties. Phys Rev Lett 48:406
43. Monserrat B (2018) Electron–phonon coupling from finite differences. J Phys Condens Matter 30
44. Jackson JD (1975) Classical electrodynamics, 2nd edn. Wiley, New York, London
45. Dai S, Fei Z, Ma Q, Rodin AS, Wagner M, McLeod AS, Liu MK, Gannett W, Regan W, Watanabe K, Taniguchi T, Thiemens M, Dominguez G, Castro Neto AH, Zettl A, Keilmann F, Jarillo-Herrero P, Fogler MM, Basov DN (2014) Tunable phonon polaritons in atomically thin van der Waals crystals of boron nitride. Science 343:1125
46. Zayats AV, Smolyaninov II, Maradudin AA (2005) Nano-optics of surface plasmon polaritons. Phys Rep 408:131
47. Lagois J, Fischer B (1982) Surface exciton polaritons from an experimental viewpoint, Chap 2. In: Mills DL, Agranovich VM (eds) Surface polaritons, modern problems in condensed matter sciences. North-Holland Pub. Co.: Sole Distributors for the USA and Canada, Elsevier Science Pub. Co., Amsterdam, New York
48. Schott M, Wegner G (1987) Basic structural and electronic properties of polydiacetylenes, Chap III-1. In: Chemla DS, Zyss J (eds) Nonlinear optical properties of organic molecules and crystals. Academic Press, p 3
49. Pierre Spinat PAR, Brouty C, Whuler A, Marie-Claude Sichi ET (1985) Determination Structurale a 205 K du Poly bis [(p-bromophenylcarbamoyloxy) -4 butyl] - 1 , 2 butene-1 yne-3 ylene , ($C_{26}H_{26}Br_2N_{204}$) n. Acta Crystallogr 41:1452
50. Enkelmann V, Schleier G (1980) Poly[1,2-bis(diphenylaminomethyl)-1-buten-3-ynylene]. Acta Crystallogr Sect B Struct Crystallogr Cryst Chem 36:1954
51. Barisien T, Legrand L, Weiser G, Deschamps J, Balog M, Boury B, Dutremez SG, Schott M (2007) Exciton spectroscopy of red polydiacetylene chains in single crystals. Chem Phys Lett 444:309

52. Schott M (2006) The colors of polydiacetylenes: a commentary. J Phys Chem B 110:15864
53. Philpott MR (1980) Optical reflection spectroscopy of organic solids. Annu Rev Phys Chem 31:97
54. Meskers SCJ, Lakhwani G (2016) Reflection of light by anisotropic molecular crystals including exciton-polaritons and spatial dispersion. J Chem Phys 145:194703
55. Tristani-Kendra M, Eckhardt CJ, Bernstein J, Goldstein E (1983) Strong coupling in the optical spectra of polymorphs of a squarylium dye. Chem Phys Lett 98:57
56. Filhol JS, Deschamps J, Dutremez SG, Boury B, Barisien T, Legrand L, Schott M (2009) Polymorphs and colors of polydiacetylenes: a first principles study. J Am Chem Soc 131:6976
57. Horvath A, Weiser G, Lapersonne-Meyer C, Schott M, Spagnoli S (1996) Wannier excitons and Franz-Keldysh effect of polydiacetylene chains diluted in their single crystal monomer matrix. Phys Rev B 53:13507
58. Wudl F, Bitler SP (1986) Synthesis and some properties of poly(diacety1ene) (polyenyne) oligomers. J Am Chem Soc 108:4685
59. Spano FC (2002) Absorption and emission in oligo-phenylene vinylene nanoaggregates: the role of disorder and structural defects. J Chem Phys 116:5877
60. Spano FC, Yamagata H (2011) Vibronic coupling in J-aggregates and beyond: a direct means of determining the exciton coherence length from the photoluminescence spectrum. J Phys Chem B 115:5133
61. Kivelson S, Heeger AJ (1985) First-order transition to a metallic state in polyacetylene: a strong-coupling polaronic metal. Phys Rev Lett 55:308
62. Giannozzi P, Baroni S, Bonini N, Calandra M, Car R, Cavazzoni C, Ceresoli D, Chiarotti GL, Cococcioni M, Dabo I, Dal Corso A, De Gironcoli S, Fabris S, Fratesi G, Gebauer R, Gerstmann U, Gougoussis C, Kokalj A, Lazzeri M, Martin-Samos L, Marzari N, Mauri F, Mazzarello R, Paolini S, Pasquarello A, Paulatto L, Sbraccia C, Scandolo S, Sclauzero G, Seitsonen AP, Smogunov A, Umari P, Wentzcovitch RM (2009) QUANTUM ESPRESSO: a modular and open-source software project for quantum simulations of materials. J Phys Condens Matter 21:395502
63. Sangalli D, Ferretti A, Miranda H, Attaccalite C, Marri I, Cannuccia E, Melo P, Marsili M, Paleari F, Marrazzo A, Prandini G, Bonfà P, Atambo MO, Affinito F, Palummo M, Molina-Sánchez A, Hogan C, Grüning M, Varsano D, Marini A (2019) Many-body perturbation theory calculations using the Yambo code. J Phys Condens Matter 31:325902
64. Mostaani E, Monserrat B, Drummond ND, Lambert CJ (2016) Quasiparticle and excitonic gaps of one-dimensional carbon chains. Phys Chem Chem Phys 18:14810

Chapter 6
Impact of Exciton Delocalisation on Exciton-Vibration Interactions

The results presented in this chapter have been published in Ref. [1] (https://doi.org/10.1103/PhysRevB.102.081122). Reprinted excerpts and figures with permission from: Antonios M. Alvertis, Raj Pandya, Loreta A. Muscarella, Nipun Sawhney, Malgorzata Nguyen, Bruno Ehrler, Akshay Rao, Richard H. Friend, Alex W. Chin, Bartomeu Monserrat. Impact of exciton delocalization on exciton-vibration interactions in organic semiconductors. Physical Review B, 102:081122(R), 2020. Copyright 2020 by the American Physical Society.

The experimental work presented in this chapter was not undertaken by the author of this thesis, but by a number of collaborators who are listed as co-authors in the aforementioned publication: Raj Pandya, Loreta A. Muscarella, Malgorzata Nguyen, Bruno Ehrler and Akshay Rao. The measurements and their interpretation are an integral part of this combined theoretical-experimental study, and are hence presented here for completeness, along with the associated experimental methodology.

Exciton-vibration interactions in organic semiconductors can become extremely strong, constituting traditional perturbative approaches for treating them inadequate. In this chapter we present a non-perturbative computational methodology that captures exciton-vibration interactions of all magnitudes. We find that the spatial delocalisation of an exciton is the key parameter determining the magnitude of its interaction with crystal versus molecular vibrations. Exciton energies computed using 'static' electronic structure calculations are renormalised once vibrational effects are accounted for, and we find that quantum nuclear motion dominates this renormalisation in most cases. We use these insights to guide experiment and reveal the full microscopic mechanism of exciton temperature- and pressure-dependence.

© The Author(s), under exclusive license to Springer Nature Switzerland AG 2021

A. M. Alvertis, *On Exciton–Vibration and Exciton–Photon Interactions in Organic Semiconductors*, Springer Theses,
https://doi.org/10.1007/978-3-030-85454-6_6

6.1 Introduction

Optoelectronic devices based on organic semiconductors, such as light-emitting diodes (LEDs) [2] and solar cells [3], are promising candidates for technological applications. In contrast to their inorganic counterparts, organic semiconductors host strongly bound excitons [4, 5] that, depending on the relative spin of the electron and hole pair, form spin-zero (singlet) or spin-one (triplet) configurations. The interconversion between singlet and triplet states is of high relevance to the application of organic semiconductors, with examples including thermally activated delayed fluorescence used in organic LEDs [6, 7], and singlet exciton fission which could lead to solar cells with efficiencies surpassing the Shockley–Queisser limit [8–10].

The weak van der Waals interaction between molecules in organic semiconductors has led to the common approximation of using isolated molecular dimers or oligomers to simulate the entire crystal [11–14]. Wavefunction-based methods are the natural way of studying such finite-sized clusters of molecules, providing a good description of the underlying physics in those systems where exciton states are localised. While triplet excitons are localised in most cases, the exchange interaction drives singlet excitons to delocalise over multiple monomers [15, 16]. Therefore, organic semiconductors can simultaneously exhibit features of the molecular and extended crystal limits, calling for a unified picture.

Due to the mechanically soft character of organic materials, the coupling of excitons to molecular and crystal vibrations can be extremely strong and dominate the physics that makes organic semiconductors interesting from an application point of view. For example, exciton-vibration coupling has been shown to play a central role in processes such as singlet fission and exciton transport [17–19], and several theoretical studies have successfully captured this interaction in molecular clusters [20–22]. However, these molecule-based approaches cannot capture the long-range intermolecular vibrations that lead to strong deviations of the orbital overlap between neighbouring monomers [23, 24]. A full treatment of exciton delocalisation becomes necessary to account for the strong dynamic disorder of organic semiconductors in solid state systems, but this remains an open challenge.

6.2 Methodology

In this chapter we present a theoretical framework to understand exciton-phonon interactions in organic semiconductors treating molecular and crystal features on the same footing. We combine many-body GW and Bethe–Salpeter calculations for exciton properties [25–27], with finite displacement methods for vibrational properties [28, 29], to capture the strong non-perturbative exciton-vibration interactions present in organic semiconductors [30]. This combination of highly accurate methods allows us to identify the extent of exciton localisation as the key parameter determining the relative importance of molecular and crystal degrees of freedom, providing a unified picture between the molecular and crystal limits of organic semiconductors. Our framework has multiple implications for the properties of these materials, some examples of which we explore. We uncover the microscopic mechanism for the weak temperature dependence of excitons in the acene series of organic semiconductors, rigorously comparing against experimental measurements. We also find that nuclear quantum fluctuations lead to strong exciton energy renormalisation, and accounting for this effect allows for unprecedented predictive power for *absolute* exciton energies compared to experiment. Finally, we predict that delocalised excitons should exhibit stronger pressure dependencies than localised ones, and again quantitatively confirm this prediction with experiments in the acene series.

At temperature T, we include the phonon contribution to the exciton energy $E_{\text{exc}}(T)$ by means of the quantum mechanical expectation value:

$$E_{\text{exc}}(T) = \frac{1}{\mathcal{Z}} \sum_{\mathbf{s}} \langle \chi_{\mathbf{s}}(\mathbf{u}) | \, E_{\text{exc}}(\mathbf{u}) \, | \chi_{\mathbf{s}}(\mathbf{u}) \rangle \, e^{-E_{\mathbf{s}}/k_{\text{B}}T}, \tag{6.1}$$

where $|\chi_{\mathbf{s}}\rangle$ is the harmonic vibrational wavefunction in state $\mathbf{s}$ with energy $E_{\mathbf{s}}$, $\mathcal{Z} = \sum_{\mathbf{s}} e^{-E_{\mathbf{s}}/k_{\text{B}}T}$ is the partition function, and $\mathbf{u}$ is a collective coordinate that includes the amplitudes of all normal modes of vibration in the system.

By substituting the vibrational wavefunction of a quantum harmonic oscillator $|\chi_{\mathbf{s}}\rangle$ we obtain the expression:

$$E_{\text{exc}}(T) = \int d\mathbf{u} |\Phi(\mathbf{u}; T)|^2 E_{\text{exc}}(\mathbf{u}), \tag{6.2}$$

where

$$|\Phi(\mathbf{u}; T)|^2 = \prod_{\mathbf{q},\nu} (2\pi \sigma^2_{\mathbf{q}\nu}(T))^{-1/2} \exp\left\{\left(-\frac{u^2_{\mathbf{q}\nu}}{2\sigma^2_{\mathbf{q}\nu}(T)}\right)\right\}, \tag{6.3}$$

the harmonic density at temperature T, which is a product of Gaussian functions of width:

$$\sigma^2_{\mathbf{q}\nu}(T) = \frac{1}{2\omega_{\mathbf{q}\nu}} \cdot \coth\left(\frac{\omega_{\mathbf{q}\nu}}{2k_B T}\right). \tag{6.4}$$

In the above, $\mathbf{q}$ is the phonon wavevector and ν is the index labelling phonon modes.

We evaluate Eq. (6.2) by generating stochastic samples distributed according to the harmonic vibrational ensemble, calculating the exciton energy and averaging over all configurations. We emphasise that while individual phonon modes are treated within the harmonic approximation, our methodology still captures exciton-phonon interactions to all orders as it does not rely on a perturbative treatment of the coupling.

To sample the single particle DFT electronic band gap using the above methodology, we generate 200 configurations, which we find lead to converged results for the calculated average. We obtain the band gap correction for each of these configurations at the three temperatures of $T = 0$ K, $T = 150$ K and $T = 300$ K. Figure 6.1 visualises these corrections for the cases of $T = 0$ K and $T = 300$ K in pentacene.

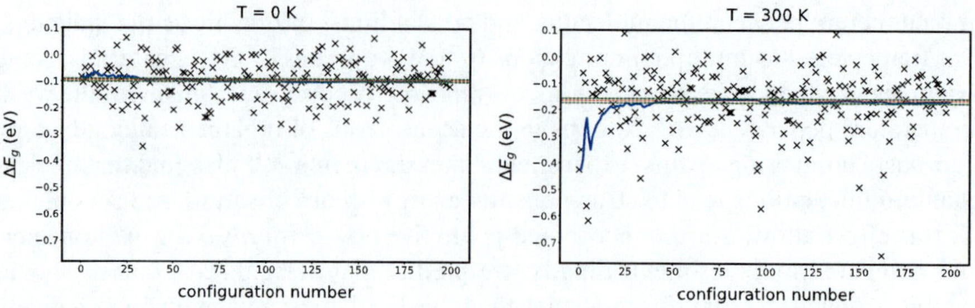

Fig. 6.1 Single particle DFT band gap correction for 200 stochastic configurations of pentacene generated at 0 K and 300 K respectively (crosses). The red lines indicate the mean band gap correction, while green lines indicate the associated statistical uncertainty. The blue lines show the running average, indicating how the band gap correction converges to the average by including more configurations. The calculated average band gap renormalisation is −96 meV at 0 K and −166 meV at 300 K

We then proceed to apply GW corrections to these DFT values. Due to the large computational cost of these calculations, we only perform GW calculations on the ten configurations whose single-particle DFT band gap value is closest to the calculated average band gap, as visualised in Fig. 6.1 for each temperature. This correlated sampling strategy between DFT and GW has been shown to be accurate in other systems [31]. Having calculated the effects of electron-phonon coupling to the quasiparticle band gap, we now solve the Bethe–Salpeter equation for the same ten configurations at the three temperatures. The parameters for all the electronic structure and phonon calculations are given in Appendix 2.

The sampling of the expectation value of Eq. 6.2 becomes increasingly accurate with the inclusion of more **q**-points in the Brillouin zone. Within the finite differences approach for phonon calculations and the expectation values of observables at finite temperatures, **q**-points are described using commensurate supercells. Due to the prohibitive computational cost of GW-BSE calculations using large supercells, for every acene crystal we study the effect of long wavelength phonons on excitons for supercells of a maximum size 2, i.e. containing two unit cells. Depending on the material, we use a $2 \times 1 \times 1$ or $1 \times 2 \times 1$ supercell, where we repeat along the dominant packing direction. This means that in our supercell calculations we include phonons at the Brillouin zone boundary, in addition to those at the Γ point that are already captured in unit cell calculations. For acene crystals (and molecular crystals in general), the weak intermolecular interactions lead to relatively flat phonon (and electron) dispersions, thus suggesting that a small number of **q**-points should be sufficient to obtain accurate Brillouin zone integrals, partly justifying the use of supercells of size 2 only. However, even supercells of size 2 are prohibitively large for performing GW calculations at the ten or more points used for each studied temperature, particularly for the larger acenes. Therefore, to obtain the phonon-renormalised exciton energies in these cases, we apply the GW correction from the unit cell calculations as a scissor operator in the supercell calculations, and then solve the BSE.

6.3 Exciton-Phonon Coupling

To probe our theoretical framework, we use the acene series of molecular crystals because they exhibit singlet excitons that range from entirely localised (molecular limit) to strongly delocalised (crystal limit), as visualised in Fig. 6.2. The electron density is shown in blue for a hole localised on the central monomer of the displayed area. Throughout the rest of this chapter, the term delocalisation will refer to the spatial extent of the electron wavefunction for a hole localised at a specific point. For singlet excitonic states, molecular size determines the degree of delocalisation in the solid state [15]. For large organic structures, the average electron-hole distance on a single molecule is comparable to the distance between neighbouring molecules. Therefore, the attractive Coulomb energy between electron and hole is similar when they localise on the same or on adjacent molecules, allowing for delocalised states with large average electron-hole separation. In contrast, in small molecules the attractive Coulomb interaction of electron and hole within a single molecule is significantly stronger than if they localised on different molecules, leading to localised singlet excitons. Triplet excitons are generally more localised than singlets, due to the lack of a repulsive exchange interaction in this state [27]. The triplet wavefunctions of the acene crystals are visualised Fig. 6.3. Therefore, depend-

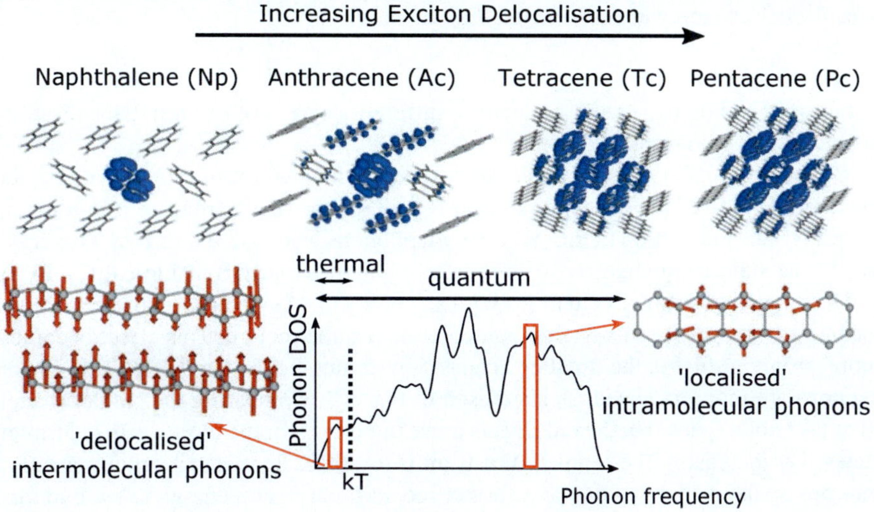

Fig. 6.2 Singlet excitons in the acene series of molecular crystals range from entirely localised (naphthalene) to highly delocalised (pentacene). The phonon density of states (DOS) of these materials consists of a thermally-activated regime of intermolecular motions, and a high-energy regime of intramolecular modes that are only active due to quantum fluctuations. The interaction of delocalised excitons is dominated by low-energy thermally-active modes, while localised excitons couple preferentially to high-energy modes via quantum fluctuations

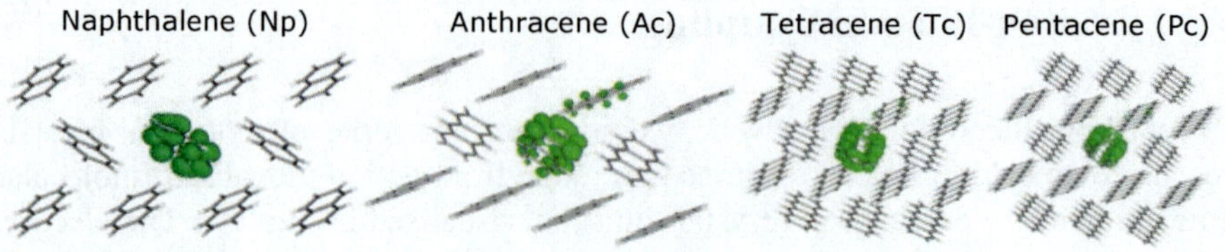

Fig. 6.3 The triplet excitons of the acene series of molecular crystals

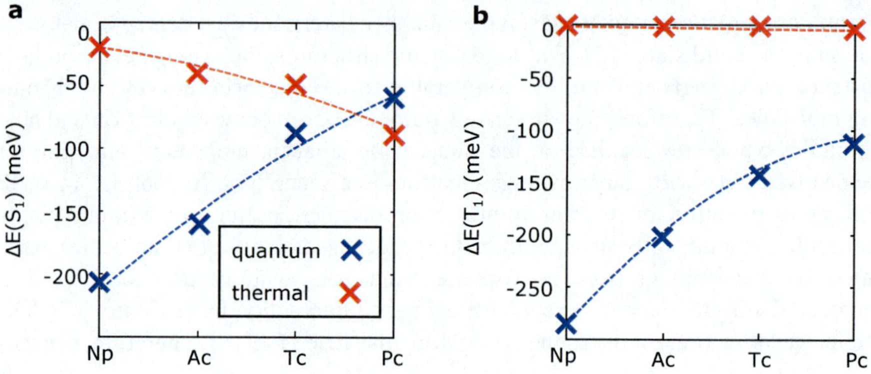

Fig. 6.4 Change caused by thermal activation (red) and quantum fluctuations (blue) of phonons to **a** the singlet and **b** triplet exciton energies of the different acenes

ing on spin and on the size of molecules, different degrees of exciton delocalisation appear in the solid state.

By employing Eq. 6.2, we obtain the expectation value for the singlet and triplet exciton energies of the acene crystals at 0 and 300 K arising from exciton-phonon coupling. The interaction of the exciton with phonons leads to a red shift of its energy, due to the stabilising character of intraband phonon-induced transitions [32]. Even at 0 K, phonon quantum fluctuations with an energy of $\frac{1}{2}\hbar\omega$ interact with the exciton, leading to a red shift of its energy compared to a static picture. This effect becomes more pronounced for the smaller acenes (fewer number of carbon rings), where excitons are more localised, as visualised in Fig. 6.4. Comparing the singlet (panel **a**) to the triplet (panel **b**), the red shift is more important for the more localised triplet states. By increasing the temperature from 0 to 300 K, low-energy intermolecular motions are activated, leading to a further red shift of exciton energies. We find this effect to be more important for more delocalised states. Indeed thermally-activated phonons have a negligible effect on the energy of the highly localised triplet states. Overall, in the molecular limit, excitons couple preferentially to high-energy phonons that are not thermally activated, so the coupling is purely via quantum fluctuations. By contrast, in the crystal limit thermally-activated low-energy phonons dominate the interaction with excitons. In intermediate cases, both contributions can be important.

To understand these results, let us consider the form of the phonon density of states (DOS) in organic semiconductors, schematically presented in Fig. 6.2 (the simulated DOS of the acene crystals is given in Fig. 6.10 of Appendix). This consists of two distinct regions: a small thermally-activated regime of low-frequency phonons, corresponding to intermolecular motion, and a large region which is dominated by high-energy intramolecular motions, such as C-C stretches. While most of the phonon modes that fall into the latter regime are not thermally-activated, they still oscillate with their zero-point energy. Thermally-activated modes that modulate the orbital overlap between neighbouring monomers [23, 24] will predominantly have an effect on more delocalised excitons, the properties of which depend on intermolecular interactions, due to their wavefunction extending over a larger number of molecules. Similarly, one expects that the high-energy intramolecular motions that are only active due to quantum fluctuations, will mostly affect localised excitons, the wavefunction of which has a greater amplitude in the vicinity of these localised motions. Therefore, based on the extent of their delocalisation, excitons in organic semiconductors respond differently to the different kinds of vibrations present in these materials. We now proceed to apply this intuitive picture summarised in Fig. 6.2 to uncover the mechanism behind the temperature-dependent properties of excitons.

6.4 Exciton Temperature Dependence

The red-shift of the exciton energy relative to its value at 0 K and due to thermally-activated phonons is visualised in Fig. 6.5 for the different acenes (dotted lines). However, exciton-phonon coupling is not enough to capture the experimentally observed temperature dependence of the exciton energy (red triangles in Fig. 6.5), and must be complemented by the effect of thermal expansion (TE—dashed lines). Thermal expansion increases intermolecular distances, thus approaching the limit of isolated single molecules, and hence entirely localised excitons. This destabilises the exciton energy, an effect which is more pronounced for delocalised singlets, and less important for states that are relatively localised to begin with. In principle it is possible to obtain the effects of thermal expansion from first principles using the quasiharmonic approximation [33], within which the free energy of the system calculated at the harmonic level is minimised as a function of volume. However, this is computationally expensive for systems like the acene crystals with several independent lattice parameters. Instead, we perform exciton calculations on a range of experimental structures obtained at different temperatures, as outlined in Sect. 2.4 of Appendix.

In all cases, the effects of thermal expansion and exciton-phonon coupling almost perfectly cancel out, leading to largely temperature-independent exciton energies (solid line), in very good agreement with experiment. While exciton delocalisation determines the magnitude of these two effects individually, the net temperature dependence of the exciton energies is overall weak. For anthracene, the inset provides a closer view to the experimental temperature dependence, highlighting the fact that the subtle interplay between exciton-phonon coupling and thermal expansion can

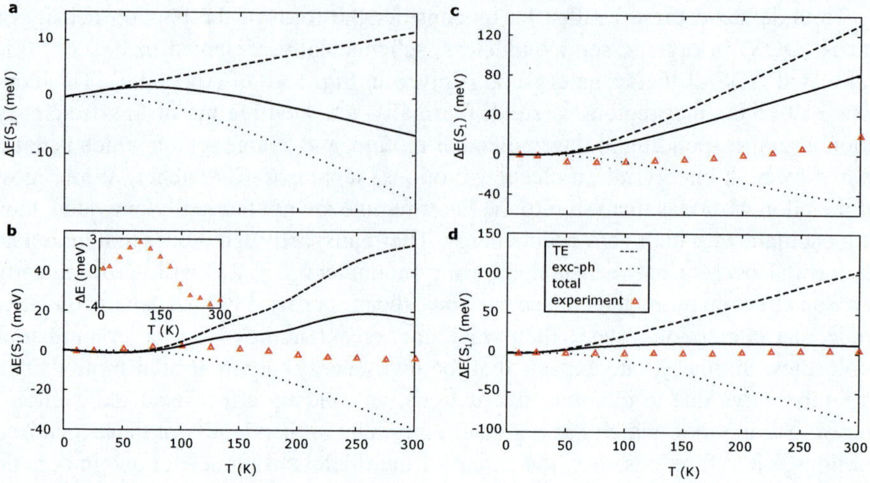

Fig. 6.5 Experimental (red-triangles) and theoretical (black solid line) temperature-dependent absorption of naphthalene (panel **a**), anthracene (panel **b**), tetracene (panel **c**) and pentacene (panel **d**). The dashed and dotted lines indicate the individual effects of thermal expansion (TE) and exciton-phonon coupling on the singlet energy

give rise to complex, non-monotonic behaviour. In this material, the slight differences between the experimental and computational temperature dependence are due to the lack of beyond zone-boundary phonons in the calculation and the anharmonicity of low-frequency phonons, which we discuss in Sect. 2.6 of Appendix. For tetracene, we overestimate the experimentally-observed blue shift of the singlet. We believe this is mainly due to two reasons:

1. Our inability to accurately account for thermal expansion, due to the lack of experimental crystal structures at a number of different temperatures. Our fit to the effect of thermal expansion is based on two structures deposited by different groups in the Cambridge Crystallographic Database. Since the two structures were also deposited in the database with 37 years of difference between them, it is possible that there are inconsistencies between the two measurements. For the rest of the acene crystals, we were able to find experimental crystal structures at a range of temperatures, which were in every case deposited within the context of the same study and by the same group of authors.
2. The lack of band edge phonons in our description of tetracene exciton-phonon coupling. As already discussed in Sect. 6.2, finite-momentum phonons can contribute significantly to the red shift of the singlet due to exciton-phonon coupling, leading to improved agreement with experiment. However, using a $1 \times 2 \times 1$ supercell for tetracene we were unable to obtain converged results within the correlated sampling strategy of 10 points described previously. Adding more points was not possible due to the very high computational cost of these calculations. We therefore restrict ourselves to the inclusion of Γ point phonons, leading to the final overestimation of the singlet blue shift.

Despite the above limitations, it is still the case that we capture the experimental trend of an increased blue shift of the tetracene singlet compared to pentacene and anthracene. The role of finite momentum phonons for the temperature dependence of tetracene and also for the rest of the acene crystals is discussed in Sect. 2.3 of Appendix.

The blue shift of exciton energies due to thermal expansion is caused by the increase in intermolecular distances, while the red shift that results from exciton-phonon coupling is due to the dominance of intraband over interband phonon-induced transitions [32]. Both these effects hold in most molecular crystals, unless one considers special cases with anomalous thermal expansion or very small optical gaps, respectively. Therefore, we expect the cancellation of the effects of thermal expansion and exciton-phonon coupling to be present in the vast majority of organic semiconductors, leading to overall weak temperature dependence of exciton energies.

6.5 The Effect of Nuclear Quantum Fluctuations on Exciton Energies

Since the effect of thermal fluctuations on exciton energies is small, it is quantum fluctuations that are mostly responsible for red shifting their values from those of a static lattice. This leads to a greatly improved agreement between theory and experiment as summarised in Fig. 6.6 and Table 6.1, highlighting the predictive power of our approach. This result is particularly important for triplet states, the energy of which is experimentally very challenging to determine [7, 34]; the correction to the static values for these states is significant due to their highly localised character and associated coupling to the high-frequency modes that are active due to quantum fluctuations. This strong renormalisation of exciton energies due to quantum fluctuations is an inherent characteristic of organic semiconductors due to the light weight of their constituent elements, and thus the high frequency of oscillations they exhibit. Therefore, accounting for this effect is crucial, regardless of the accuracy of the underlying electronic structure method. The accurate calculation of exciton energies is crucial to various photophysical processes in organic semiconductors. For the example of singlet fission, we successfully predict the experimentally well-established exothermicity ($E(S_1) > 2 \cdot E(T_1)$) of singlet fission in solid pentacene [35], which is not captured by many-body perturbation theory in the absence of quantum fluctuations [16, 36]. We also capture the endothermicity of singlet fission in tetracene [37].

6.6 Exciton Pressure Dependence

The magnitude of the effect of thermal expansion and exciton-phonon coupling on exciton energies depends on the degree of wavefunction delocalisation. Unfortu-

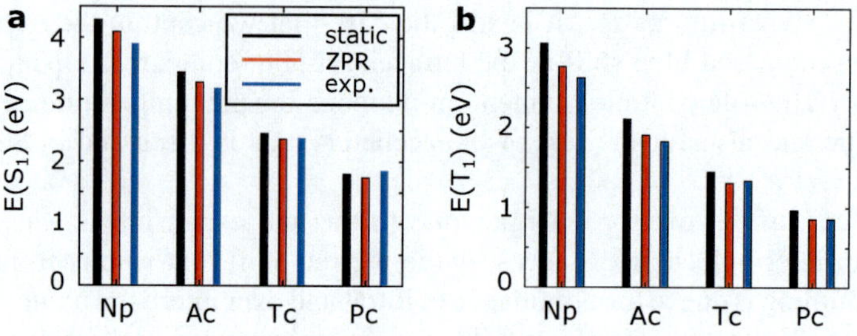

Fig. 6.6 Comparison of the computed static exciton energies obtained to their zero-point renormalised (ZPR) values and to experiment

Table 6.1 Quantum fluctuations lead to a zero-point renormalisation (ZPR) of the exciton energies calculated at the static lattice level, leading to a much improved agreement with experiment

	Calculated static	ZPR (meV)	Calculated with phonons	Experiment
$E(S_1)$ (eV)				
Naphthalene	4.28	−190	4.09	3.90 [38]
Anthracene	3.44	−160	3.28	3.11 −3.26 [38]
Tetracene	2.45	−90	2.36	2.36 [39] −2.39
Pentacene	1.80	−60	1.74	1.83 [35] −1.88
$E(T_1)$ (eV)				
Naphthalene	3.06	−290	2.77	2.63 [40]
Anthracene	2.11	−200	1.91	1.83 [7]
Tetracene	1.45	−140	1.31	1.34 [7]
Pentacene	0.97	−110	0.86	0.86 [41]

For the singlet energy of anthracene, tetracene and pentacene, we present a range of values, based both on literature values and measurements performed in the context of the study on which this chapter is based [1] (Sect. 1.1 of Appendix for experimental spectra)

nately, the cancellation of these effects that we have demonstrated does not allow one to utilise them as a means of probing delocalisation experimentally. Nevertheless, the insights from the above discussion motivate us to examine the effect of pressure on exciton energies. The application of hydrostatic pressure at a given temperature reduces the volume V_o of the unit cell at atmospheric pressure ($\Delta V/V_o < 0$), an effect *opposite* to thermal expansion, which increases the unit cell volume ($\Delta V/V_o > 0$). However, unlike thermal expansion, the effect of pressure does not compete with phonon-activated processes, leading to a strong red shift of exciton energies [42], an effect that we expect to be stronger for more delocalised states, and could hence be used to probe delocalisation. For obtaining the exciton properties at finite pressures, we relax both the internal structure and volume of the unit cell, minimising the enthalpy at a given pressure. For each acene crystal, we compute the exciton properties at 0, 0.5, 1, 2, 4 and 5 GPa. We also account for the change in the effect

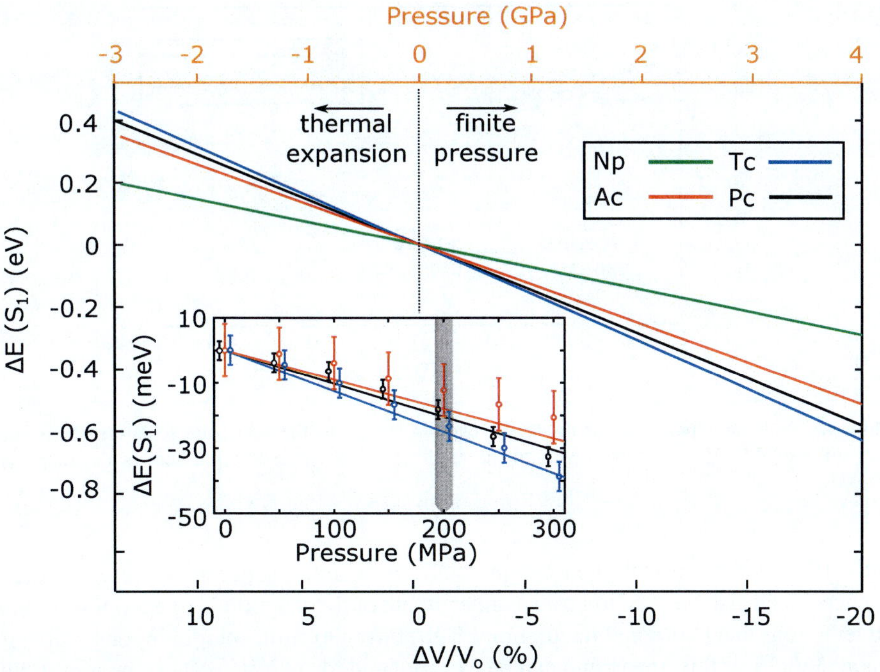

Fig. 6.7 Theoretical volume and pressure dependence of exciton energies. We compare to pressure-dependent measurements within 0–300 MPa in the inset

of phonon quantum fluctuations under pressure, and Sect. 2.5 of Appendix provides details on the importance of this effect.

Figure 6.7 shows the change of the singlet energy of the acene crystals as a function of unit cell volume. As argued previously, changes in intermolecular distances induced by pressure or thermal expansion have a stronger effect on delocalised states. Indeed, we find that for naphthalene the singlet volume dependence is weaker than for anthracene, which in turn is weaker than that of tetracene. Interestingly, pentacene has a slightly weaker volume dependence than tetracene, which is due to qualitative changes in the effect of phonon quantum fluctuations at finite pressures. Consistent with this picture, we find in Fig. 6.8 that the pressure dependence of the triplet energies is significantly weaker, due to the highly localised nature of these states. In the inset of Fig. 6.7, we compare our theoretical predictions for the change of the singlet energies at finite pressure to experiment, in the range of 0–300 MPa. The full experimental spectra are given in Fig. 6.9 of Appendix. We find that theory and experiment are in very good agreement and, remarkably, we correctly predicted the unconventional pressure dependence for pentacene, confirming the accuracy of our theoretical framework.

The above discussion shows that the slope of the experimental exciton energy pressure dependence provides a qualitative measure of exciton delocalisation. Addition-

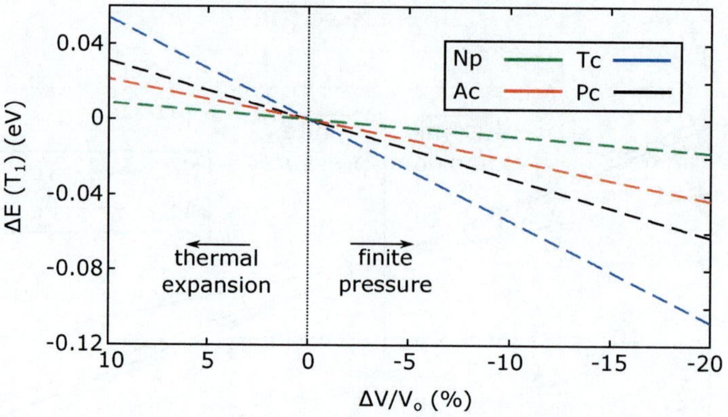

Fig. 6.8 Volume dependence of the triplet exciton energy of the acene molecular crystals. The region with $\Delta V/V_o < 0$ corresponds to the application of hydrostatic pressure in the range of 0–4 GPa

ally, in combination with the experimentally measured exciton temperature dependence, it can also be used to provide an estimate of the magnitude of exciton-phonon interactions due to thermal fluctuations. The shift of exciton energies upon compression $\Delta V/V_o < 0$ is linear, and can be extrapolated to $\Delta V/V_o > 0$, hence providing an expected blue shift of the exciton energy due to thermal expansion. The difference of this expected blue shift from the experimentally measured energy shift (compared to 0 K) gives the magnitude of exciton energy renormalisation due to coupling to thermally-activated low-frequency phonons. This is complementary to extracting the magnitude of exciton-phonon interactions from the vibronic progression of the absorption spectrum [43], as the latter only provides information on the coupling of the exciton to high-frequency modes.

6.7 Conclusions

We propose a general framework to study exciton-phonon interactions in organic semiconductors describing localised molecular and extended crystal degrees of freedom simultaneously. We show that exciton delocalisation determines the magnitude and nature of these interactions: localised excitons predominantly couple to high-frequency modes via quantum fluctuations, while delocalised excitons interact more strongly with thermally-activated low-frequency phonons. Together with the effect of thermal expansion, which also depends on exciton delocalisation, this allows us to reveal the full microscopic mechanism behind the weak temperature of exciton energies in acene crystals, and argue that this should hold in the vast majority of molecular crystals. As a consequence of the weak temperature dependence, the major contribution to exciton energy renormalisation compared to the static lattice arises from quantum fluctuations of mostly high-frequency vibrations, always present in organic materials. The magnitude of this renormalisation also depends sensitively on exciton delocalisation, and accounting for this effect is necessary in order to achieve predictive power for exciton energies.

Overall, our framework provides a unifying picture between the molecular and crystal limits of organic semiconductors, showing how the delocalisation of excitons determines their response to a wide range of structural changes, beyond lattice vibrations and thermal expansion. The effect of pressure provides such an example, and we find that pressure-dependent measurements may be used as a probe of delocalisation and thermally-activated exciton-phonon interactions. Therefore, based on factors that determine exciton delocalisation, such as spin and the size of molecules, one can anticipate the difference in the response of different materials to a variety of structural changes.

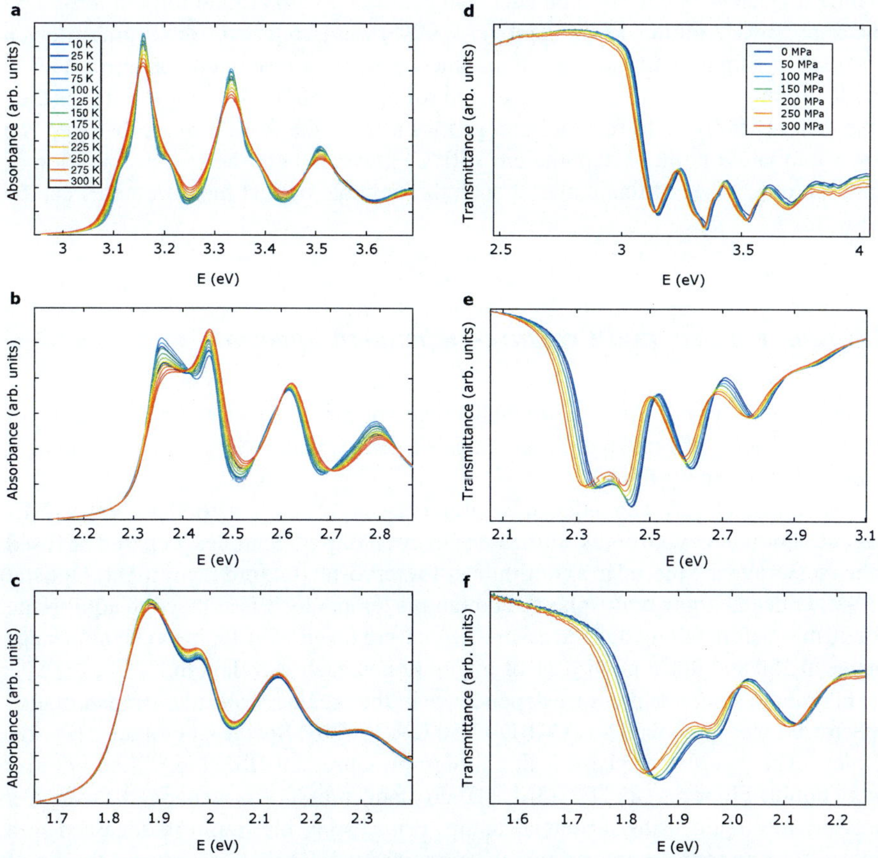

Fig. 6.9 Experimental temperature dependent absorption spectra of anthracene, tetracene and pentacene (panels **a**, **b**, **c** respectively), and pressure-dependent transmittance (panels **d**, **e**, **f** respectively)

Appendix 1: Experimental Details

While the experimental work presented in this chapter was not undertaken by the author of this thesis, the relevant details are given here for completeness.

1.1 Sample Preparation

Tetracene and pentacene powders were bought from Sigma Aldrich (99% purity) and used as supplied. The powder is transferred to an evaporator chamber (Kurt J Lesker) and kept overnight under high vacuum ($< 10^{-6}$ mbar) to remove any residual solvent prior to thermal evaporation. A 150-nm thick layer of pentacene/tetracene is then thermally evaporated at a rate of 0.5 ± 0.1 Å/s. The rate is monitored using a standard Quartz-crystal microbalance and calibrated with atomic force microscopy measurements. Anthracene (reagent grade 97%) was purchased from Sigma Aldrich and used as supplied. Crystallites of anthracene are dissolved in choloroform (HPLC grade, Sigma Aldrich) to make a saturated solution at 50 °C. The hot solution is then spin coat at 1000 rpm to form a homogenous film on the fused silica substrate. The low sublimation point of naphthalene (80 °C) prevents both thermal evaporation or solution processing of this material such that reliable optical measurements can be performed.

1.2 Temperature- and Pressure-Dependent Spectra

Figure 6.9 summarises the experimental temperature-dependent absorption spectra of anthracene, tetracene and pentacene (panels **a–c**), as well as the pressure-dependent transmittance (panels **d–f**).

For all temperature-dependent measurements, an Agilent Cary 6000i UV–vis–NIR spectrophotometer with blank substrate correction is used. Samples prepared on fused silica substrates are placed in a continuous-flow cryostat (Oxford Instruments Optistat CF-V) under an argon atmosphere. The sample temperature is allowed to equilibrate for 30 min before taking data. Measurements were taken with 1 nm wavelength steps between 250 and 900 nm, with 1 s of integration at each wavelength.

In order to study the pressure dependence of the organic molecules, transmittance spectra are measured with a LAMBDA 750 UV/Vis/NIR Spectrophotometer (Perkin Elmer). The samples are kept inside a high-pressure cell (ISS Inc.) filled with an inert liquid, Fluorinert FC-72 (3M). Hydrostatic pressure is generated through a pressurising liquid using a manual pump. Prior using, the liquid is degassed in a Schlenk line to remove oxygen which causes, from 300 MPa onwards, scattering of a fraction of light and therefore a reduction of the transmitted signal from the sample. The pressure is applied from ambient pressure to 300 MPa in steps of 50 MPa. Before the measurement, 7 min are allowed for equilibration of the material under pressure. The error of the pressure reading is estimated to be 20 MPa.

Appendix 2: Computational Details

2.1 DFT and GW-BSE Calculations

We perform DFT calculations using the Quantum Espresso software package [44], with the PBE functional together with a 60 Ry plane wave cutoff energy. For pentacene and tetracene, we employ a $4 \times 4 \times 2$ **k**-point grid, while for anthracene and naphthalene, we use $4 \times 4 \times 4$ and $2 \times 4 \times 2$ grids respectively. The static lattice values of the acene exciton energies reported in Fig. 6.6 and Table 6.1 of the main text correspond to the structures PENCEN08, TETCEN03, ANTCEN19 and NAPHTA36 respectively, as deposited in the Cambridge Crystallographic Database [45–48]. Prior to any electronic structure and/or phonon calculation, we relax the internal coordinates of the unit cells, leaving their volume fixed, and using the Tkatchenko–Scheffler van der Waals dispersion correction [49].

We perform energy self-consistent GW calculations (i.e. GW_0) using the YAMBO code [50]. We find that including 300 Kohn–Sham states, 200 bands for the calculation of the polarisation function, and a 7 Ry cutoff for the dielectric matrix, leads to converged values for the quasiparticle band gaps. For the singlet and triplet excitonic states, we solve the Bethe–Salpeter equation (BSE) using 8 occupied and 8 unoccupied bands, converging the position of the first exciton peak. For the exchange term in the Bethe–Salpeter kernel, the cutoff is set to 40 Ry for singlet calculations, while it is set to zero for triplet calculations.

2.2 Phonon Calculations

We calculate the lattice dynamics of the acene crystals in the harmonic approximation using finite differences [28] in conjunction with nondiagonal supercells [51]. We calculate the dynamical matrix on a $2 \times 1 \times 1$ coarse **q**-point grid for pentacene, and $1 \times 2 \times 1$ grids for tetracene, anthracene, and naphthalene. We then use Fourier interpolation on 10^5 stochastically generated **q**-points to construct the phonon density of states (DOS), shown in Fig. 6.10a. There are also a few phonon modes corresponding to C–H stretches in the region 3100–3150 cm^{-1}, which we have excluded from the plot for visibility reasons. Figure 6.10b highlights the low-frequency modes of the acene crystals.

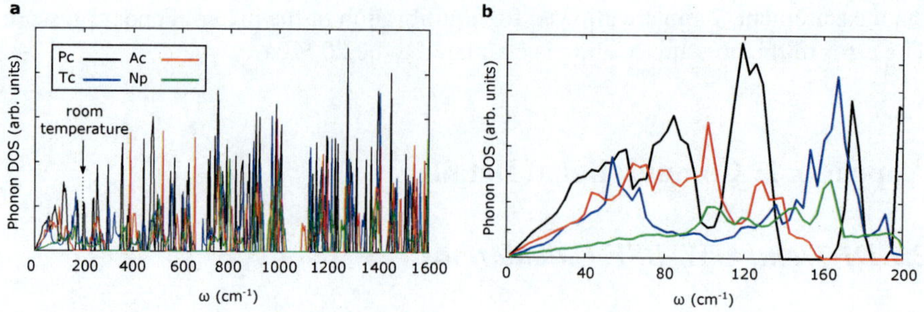

Fig. 6.10 Phonon density of states (DOS) of the acene crystals. We mark room temperature, indicating that the modes below that energy can be thermally activated within the studied range of temperatures

2.3 Phonon q-Point Sampling

Table 6.2 summarises the change in the exciton energies when increasing the temperature from 0 to 300 K, due to coupling with phonons. Results are given for the two cases of including Γ point phonons only, and for including Γ along with band-edge phonons. For tetracene, we were unable to obtain converged results within a computationally feasible number of displaced configurations.

For the delocalised singlet states of anthracene and pentacene, inclusion of band edge phonons leads to a stronger red shift of their energy compared to only including Γ point phonons. This is very important for capturing the experimental behaviour of Fig. 3 in the main text, as the mere inclusion of Γ phonons would suggest that the singlet energy blue shifts substantially. This is also the reason that we overestimate the temperature-induced blue shift of the tetracene singlet in Fig. 6.5. Furthermore, it is worth noting that long-wavelength phonons that are captured through supercell calculations mostly have a weak effect on the energy of the localised triplet states, as expected. Naphthalene shows a distinct behaviour than the rest of the acenes, with band edge phonons resulting in an overall weak renormalisation of its exciton energies, which is not surprising given their highly localised character.

2.4 Thermal Expansion Calculations

In order to study the effects of thermal expansion, we perform DFT and GW-BSE calculations on a range of structures at different temperatures, deposited on the Cambridge Crystallographic Database. The references to each structure are given in Table 6.3. Prior to the electronic structure calculations, we relax the internal coordinates of the unit cell, while keeping the volume of the cell constant. For the optimisation, we employ a $4 \times 4 \times 2$ **k**-point grid and use the Tkatchenko–Scheffler van der Waals corrections [49]. In some cases, and particularly for tetracene, it was

Table 6.2 The effect of temperature-activated phonons on exciton energies, in the cases of including Γ phonons only, and including Γ phonons as well as phonons at the boundary of the Brillouin zone

	Γ phonons only (meV)	Γ and zone-boundary phonons (meV)
$\Delta E(S_1) = E(S_1)_{300\,K} - E(S_1)_{0\,K}$		
Naphthalene	−46	−19
Anthracene	−21	−39
Tetracene	−52	–
Pentacene	−41	−86
$\Delta E(T_1) = E(T_1)_{300\,K} - E(T_1)_{0\,K}$		
Naphthalene	−46	−0
Anthracene	−5	−1
Tetracene	−2	–
Pentacene	−2	−3

Table 6.3 Crystal structures of acenes used in order to study the effects of thermal expansion, along with the temperatures at which they were measured and the Cambridge Crystallographic Database reference

Material	Crystal structure references	Temperature (K)
Naphthalene [48]	NAPHTA32	30
	NAPHTA34	150
	NAPHTA30	220
	NAPHTA36	295
Anthracene [52]	ANTCEN09	94
	ANTCEN11	181
	ANTCEN12	220
	ANTCEN13	259
	ANTCEN14	295
Tetracene [53, 54]	TETCEN01	175
	TETCEN	295
Pentacene [45]	PENCEN06	120
	PENCEN07	293
	PENCEN08	414

unfortunately only possible to find a limited number of crystal structures at different temperatures, limiting our ability to accurately describe the effects of thermal expansion over the whole range of studied temperatures.

2.5 Phonon Quantum Fluctuations and the Effect of Pressure

When applying hydrostatic pressure to a molecular crystal, the material becomes stiffer, and the phonon energies increase [42]. While this alone would suggest that zero-point fluctuations leads to increasingly stronger exciton energy renormalisations at large pressures compared to ambient conditions, increasing pressure also leads to a decrease in intermolecular distance, allowing the electronic part of the wavefunction of an exciton state to delocalise further away from the position of the hole. Since zero-point fluctuations have a more prominent effect on localised states, this argument suggests that their effect becomes weaker at large pressures.

Therefore, under hydrostatic pressure, the increase in phonon frequencies and in exciton delocalisation compete in the contribution of phonon quantum fluctuations on exciton energies. In order to gain a better understanding of this, we calculate the phonon modes of the acene crystals at 4 GPa and then use our computational methodology to quantify the effect of quantum fluctuations on exciton energies at this pressure. We compare the zero-point renormalisation (ZPR) of exciton energies between the cases of atmospheric pressure and 4 GPa in Table 6.4.

The change in the effect of quantum fluctuations on exciton energies seems to depend sensitively on the subtle details of individual compounds. For most exciton states of the acene crystals, the effect seems to become weaker at larger pressures, indicating that increased exciton delocalisation is more important than the increase in phonon frequencies. However, the anthracene singlet state seems to provide an exception to this. The most important change is found for pentacene, for which quantum fluctuations lead to a blue shift under the application of finite hydrostatic pressure. This blue shift of the pentacene singlet energy competes with the red shift that increasing exciton delocalisation produces, hence the overall red shift becomes weaker, which is why for both the singlet and triplet states of pentacene (Fig. 5 in main text and Fig. 6.8) we find a weaker pressure/volume dependence than for tetracene, despite the pentacene excitons being more delocalised.

For all the studied materials, the slope of the exciton volume/pressure dependence is strongly influenced by the variation in the effect of quantum fluctuations at finite pressures. Table 6.5 gives the slope of the pressure-dependent absorption of the acene crystals with and without accounting for this effect, compared to the measured experimental values. Ignoring the variation in quantum fluctuations results in overestimating the pressure dependence, while including this effect significantly improves the agreement to experiment, also correctly exchanging the order between the tetracene and pentacene values.

2.6 Anharmonicity of Low-Frequency Modes

The discussion in Sect. 2.3 of Appendix highlights that accounting for finite momentum phonons can lead to a stronger red shift of exciton energies, particularly for more

Table 6.4 Zero-point renormalisation (ZPR) of the first excited singlet and triplet state of the acene crystals, without the application of hydrostatic pressure, and with a hydrostatic pressure of 4 GPa

	ZPR at 0 GPa (meV)	ZPR at 4 GPa (meV)
S_1		
Naphthalene	−190	−188
Anthracene	−157	−195
Tetracene	−82	−54
Pentacene	−57	+307
T_1		
Naphthalene	−292	−257
Anthracene	−200	−191
Tetracene	−132	−58
Pentacene	−116	+214

Table 6.5 Slope of the pressure-dependent singlet energy of the acene crystals, with and without including the variation of zero-point quantum renormalisation (ZPR) of exciton energies at finite pressures, and comparison to experiment

Slope of $E(S_1)$ versus P	Without ZPR variation (meV/GPa)	With ZPR variation (meV/GPa)	Experiment (meV/GPa)
Naphthalene	−82	−75	–
Anthracene	−102	−90	−73
Tetracene	−138	−126	−133
Pentacene	−201	−102	−103

delocalised states. For the case of anthracene, the computed temperature dependence of the singlet energy slightly overestimates the experimental trend, as presented in Fig. 6.5b, which we argue is due to the lack of finite momentum phonons beyond the zone boundary. Given the relatively small size of the anthracene primitive cell, we attempted to test this by using our computational methodology with a finer **q**-point sampling grid including 4 points (in reciprocal lattice relative coordinates): (0, 0, 0), (0, 0, 0.5), (0, 0.5, 0), (0, 0.5, 0.5), corresponding to a calculation on a supercell of size $1 \times 2 \times 2$. Unfortunately, this test was unsuccessful but highlighted another area in which our computational framework could be improved in the future. The predicted temperature-induced red shift of the exciton energies due to coupling with phonons is −260 meV, which is unphysically large compared to the anthracene singlet energy corrections that are summarised in Table 6.2. This is due to the fact that in the larger supercell, phonon modes of very low frequencies appear, and in these modes the harmonic approximation is no longer valid. To demonstrate this point, we plot in Fig. 6.11a the potential energy surface (PES) of the lowest frequency phonon for a $1 \times 2 \times 2$ supercell calculation (black), along with the PES of the same mode within the harmonic approximation (red). It is evident that the

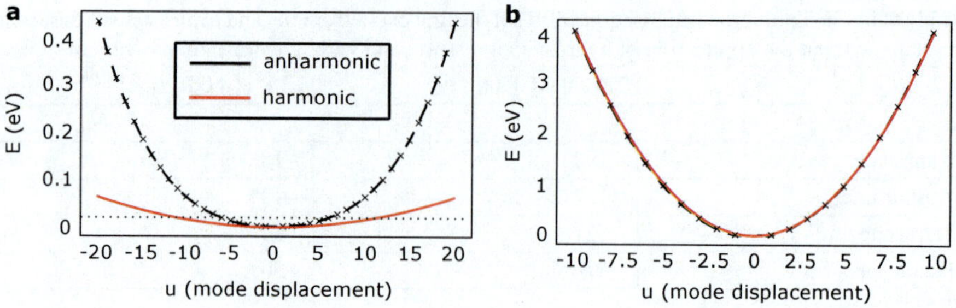

Fig. 6.11 Comparison of the *ab initio* potential energy surface to the harmonic approximation, for a low-frequency (panel **a**), and a high-frequency mode (panel **b**). Room temperature in panel **a** is indicated by a dotted line

harmonic approximation breaks down in this case. In contrast to that, the harmonic approximation works very well for a high-frequency mode in Fig. 6.11b, as expected.

At a given finite temperature, phonon modes are populated according to Bose–Einstein statistics, hence low-frequency modes are strongly displaced. However, it becomes evident from Fig. 6.11a that using the harmonic approximation reduces the potential barrier at room temperature, leading to significantly stronger displacements of these modes compared to the case where higher terms are included. This leads to unphysically distorted structures and results in an overestimation of the exciton-phonon interaction strength. It will be interesting to explore the role of anharmonic vibrations on exciton-phonon coupling in future work.

References

1. Alvertis AM, Pandya R, Muscarella LA, Sawhney N, Nguyen M, Ehrler B, Rao A, Friend RH, Chin AW, Monserrat B (2020) Impact of exciton delocalization on exciton-vibration interactions in organic semiconductors. Phys Rev B 102:081122(R)
2. Reineke S, Lindner F, Schwartz G, Seidler N, Walzer K, Lüssem B, Leo K (2009) White organic light-emitting diodes with fluorescent tube efficiency. Nature 459:234
3. Meng L, Zhang Y, Wan X, Li C, Zhang X, Wang Y, Ke X, Xiao Z, Ding L, Xia R, Yip H-L, Cao Y, Chen Y (2018) Organic and solution-processed tandem solar cells with 17.3% efficiency. Science 361:1094
4. Knupfer M (2003) Exciton binding energies in organic semiconductors. Appl Phys A 77:623
5. Köhler A, Bässler H (2009) Triplet states in organic semiconductors. Mater Sci Eng R Rep 66:71
6. Uoyama H, Goushi K, Shizu K, Nomura H, Adachi C (2012) Highly efficient organic light-emitting diodes from delayed fluorescence. Nature 492:234
7. Reineke S, Baldo MA (2014) Room temperature triplet state spectroscopy of organic semiconductors. Sci Rep 4:3797
8. Smith MB, Michl J (2010) Singlet fission. Chem Rev 110:6891
9. Hanna MC, Nozik AJ (2006) Solar conversion efficiency of photovoltaic and photoelectrolysis cells with carrier multiplication absorbers. J Appl Phys 100:074510
10. Rao A, Friend RH (2017) Harnessing singlet exciton fission to break the Shockley–Queisser limit. Nat Rev Mater 2:17063

11. Beljonne D, Yamagata H, Brédas JL, Spano FC, Olivier Y (2013) Charge-transfer excitations steer the Davydov splitting and mediate singlet exciton fission in pentacene. Phys Rev Lett 110:226402
12. Berkelbach TC, Hybertsen MS, Reichman DR (2013) Microscopic theory of singlet exciton fission. II. Application to pentacene dimers and the role of superexchange. J Chem Phys 138:114103
13. Yost SR, Lee J, Wilson MWB, Wu T, McMahon DP, Parkhurst RR, Thompson NJ, Congreve DN, Rao A, Johnson K, Sfeir MY, Bawendi MG, Swager TM, Friend RH, Baldo M, Van Voorhis T (2014) A transferable model for singlet-fission kinetics. Nat Chem 6:492
14. Coto PB, Sharifzadeh S, Neaton JB, Thoss M (2015) Low-lying electronic excited states of pentacene oligomers: a comparative electronic structure study in the context of singlet fission. J Chem Theory Comput 11:147
15. Cudazzo P, Sottile F, Rubio A, Gatti M (2015) Exciton dispersion in molecular solids. J Phys Condens Matter 27
16. Refaely-Abramson S, Da Jornada FH, Louie SG, Neaton JB (2017) Origins of singlet fission in solid pentacene from an ab initio Green's function approach. Phys Rev Lett 119:267401
17. Aragó J, Troisi A (2015) Dynamics of the excitonic coupling in organic crystals. Phys Rev Lett 114:026402
18. Musser AJ, Liebel M, Schnedermann C, Wende T, Kehoe TB, Rao A, Kukura P (2015) Evidence for conical intersection dynamics mediating ultrafast singlet exciton fission. Nat Phys 11:352
19. Bakulin AA, Morgan SE, Kehoe TB, Wilson MWB, Chin AW, Zigmantas D, Egorova D, Rao A (2016) Real-time observation of multiexcitonic states in ultrafast singlet fission using coherent 2D electronic spectroscopy. Nat Chem 8:16
20. Miyata K, Kurashige Y, Watanabe K, Sugimoto T, Takahashi S, Tanaka S, Takeya J, Yanai T, Matsumoto Y (2017) Coherent singlet fission activated by symmetry breaking. Nat Chem 9:983
21. Schnedermann C, Alvertis AM, Wende T, Lukman S, Feng J, Schröder FAYN, Turban DHP, Wu J, Hine NDM, Greenham NC, Chin AW, Rao A, Kukura P, Musser AJ (2019) A molecular movie of ultrafast singlet fission. Nat Commun 10:4207
22. Alvertis AM, Schröder FAYN, Chin AW (2019) Non-equilibrium relaxation of hot states in organic semiconductors: impact of mode-selective excitation on charge transfer. J Chem Phys 151:084104
23. Troisi A, Orlandi G, Anthony JE (2005) Electronic interactions and thermal disorder in molecular crystals containing cofacial pentacene units. Chem Mater 17:5024
24. Troisi A, Orlandi G (2006) Dynamics of the intermolecular transfer integral in crystalline organic semiconductors. J Phys Chem A 110:4065
25. Lars H (1965) New method for calculating the one-particle Green's function with application to the electron-gas problem. Phys Rev 139:796
26. Rohlfing M, Louie SG (1998) Electron-hole excitations in semiconductors and insulators. Phys Rev Lett 81:2312
27. Rohlfing M, Louie SG (2000) Electron-hole excitations and optical spectra from first principles. Phys Rev B 62:4927
28. Kunc K, Martin RM (1982) Ab initio force constants of GaAs: a new approach to calculation of phonons and dielectric properties. Phys Rev Lett 48:406
29. Monserrat B (2018) Electron–phonon coupling from finite differences. J Phys Condens Matter 30
30. Monserrat B (2016) Vibrational averages along thermal lines. Phys Rev B 93:014302
31. Monserrat B (2016) Correlation effects on electron-phonon coupling in semiconductors: many-body theory along thermal lines. Phys Rev B 93:100301(R)
32. Saha K, Garate I (2014) Phonon-induced topological insulation. Phys Rev B 89:205103
33. Mounet N, Marzari N (2005) First-principles determination of the structural, vibrational and thermodynamic properties of diamond, graphite, and derivatives. Phys Rev B 71:205214

34. Ehrler B, Walker BJ, Böhm ML, Wilson MWB, Vaynzof Y, Friend RH, Greenham NC (2012) In situ measurement of exciton energy in hybrid singlet-fission solar cells. Nat Commun 3:1019
35. Rao A, Wilson MWB, Albert-Seifried S, Di Pietro R, Friend RH (2011) Photophysics of pentacene thin films: the role of exciton fission and heating effects. Phys Rev B 84:195411
36. Liu X, Tom R, Wang X, Cook C, Schnatschneider B, Marom N (2020) Pyrene-stabilized acenes as intermolecular singlet fission candidates: importance of exciton wave-function convergence. J Phys Condens Matter 32:184001
37. Burdett JJ, Bardeen CJ (2013) The dynamics of singlet fission in crystalline tetracene and covalent analogs. Acc Chem Res 46:1312
38. Schnepp O (1963) Electronic spectra of molecular crystals. Annu Rev Phys Chem 14:35
39. Wilson MWB, Rao A, Johnson K, Gélinas S, Di Pietro R, Clark J, Friend RH (2013) Temperature-independent singlet exciton fission in tetracene. J Am Chem Soc 135:16680
40. Swiderek P, Michaud M, Hohlneicher G, Sanche L (1990) Electron energy loss spectroscopy of solid naphthalene and acenaphthene: search for the low-lying triplet states. Chem Phys Lett 175:667
41. Rao A, Wilson MWB, Hodgkiss JM, Albert-Seifried S, Bässler H, Friend RH (2010) Exciton fission and charge generation via triplet excitons in pentacene/C_{60} bilayers. J Am Chem Soc 132:12698
42. Farina L, Syassen K, Brillante A, Della Valle RG, Venuti E, Karl N (2003) Pentacene at high pressure. High Press Res 23:349
43. Spano FC (2010) The spectral signatures of Frenkel polarons in H- and J-aggregates. Acc Chem Res 43:429
44. Giannozzi P, Baroni S, Bonini N, Calandra M, Car R, Cavazzoni C, Ceresoli D, Chiarotti GL, Cococcioni M, Dabo I, Dal Corso A, De Gironcoli S, Fabris S, Fratesi G, Gebauer R, Gerstmann U, Gougoussis C, Kokalj A, Lazzeri M, Martin-Samos L, Marzari N, Mauri F, Mazzarello R, Paolini S, Pasquarello A, Paulatto L, Sbraccia C, Scandolo S, Sclauzero G, Seitsonen AP, Smogunov A, Umari P, Wentzcovitch MR (2009) QUANTUM ESPRESSO: a modular and open-source software project for quantum simulations of materials. J Phys Condens Matter 21:395502
45. Haas S, Batlogg B, Besnard C, Schiltz M, Kloc C, Siegrist T (2007) Large uniaxial negative thermal expansion in pentacene due to steric hindrance. Phys Rev B 76:205203
46. Robertson JM, Sinclair VC, Trotter J (1961) The crystal and molecular structure of tetracene. Acta Crystallogr 14:697
47. Malecki JG (2013) CSD Commun. CCDC 950158
48. Capelli SC, Albinati A, Mason SA, Willis BTM (2006) Molecular motion in crystalline naphthalene: analysis of multi-temperature X-ray and neutron diffraction data. J Phys Chem A 110:11695
49. Tkatchenko A, Scheffler M (2009) Accurate molecular van der Waals interactions from ground-state electron density and free-atom reference data. Phys Rev Lett 102:073005
50. Sangalli D, Ferretti A, Miranda H, Attaccalite C, Marri I, Cannuccia E, Melo P, Marsili M, Paleari F, Marrazzo A, Prandini G, Bonfà P, Atambo MO, Affinito F, Palummo M, Molina-Sánchez A, Hogan C, Grüning M, Varsano D, Marini A (2019) Many-body perturbation theory calculations using the yambo code. J Phys Condens Matter 31:325902
51. Lloyd-Williams JH, Monserrat B (2015) Lattice dynamics and electron-phonon coupling calculations using nondiagonal supercells. Phys Rev B 92:184301
52. Brock CP, Dunitz JD (1990) Temperature dependence of thermal motion in crystalline anthracene. Acta Crystallogr Sect B 46:795
53. Holmes D, Kumaraswamy S, Matzger AJ, Vollhardt KPC (1999) On the nature of nonplanarity in the [N]phenylenes. Chem Eur J 5:3399
54. Campbell RB, Robertson JM, Trotter J (1962) The crystal structure of hexacene, and a revision of the crystallographic data for tetracene. Acta Crystallogr 15:289

Chapter 7
Interplay of Vibrational Relaxation and Charge Transfer

The results presented in this chapter have been published in Ref. [1] (https://doi.org/10.1063/1.5115239), where parts of the text and figures appear. Reprinted from:

Antonios M. Alvertis, Florian A.Y.N. Schröder, Alex W. Chin. Non-equilibrium relaxation of hot states in organic semiconductors: impact of mode-selective excitation on charge transfer. The Journal of Chemical Physics, 151:084104, 2019, with the permission of AIP Publishing.

Having studied the interaction of excitons with photons and vibrations in equilibrium in Chaps. 5 and 6 respectively, we now proceed to study non-equilibrium exciton dynamics, as they arise following light-absorption. For the picosecond timescales we consider it is safe to work in the weak exciton-photon coupling regime, i.e. assuming absorption to be a 'one-off' event, which generates a pure exciton state. As discussed in the introduction to the thesis and Chap. 3, exciton-vibration interactions can induce non-adiabatic transitions, leading to a breakdown of the Born-Oppenheimer approximation. Therefore, we utilise a beyond Born-Oppenheimer method based on tensor networks in order to study exciton-vibration interactions following light-absorption from a covalent tetracene dimer. We study the process of endothermic intramolecular charge transfer and show that excess energy excitation can enhance its efficiency. We find the specific pathway of vibrational relaxation to have an important effect on the final charge transfer yield, and we discuss the qualitative differences between low- and high-frequency vibrations in facilitating ultrafast dynamics.

7.1 Introduction

Light-harvesting in organic molecules is a fundamental process which has attracted a lot of attention in recent years, its applications ranging from photosynthetic sys-

© The Author(s), under exclusive license to Springer Nature Switzerland AG 2021
A. M. Alvertis, *On Exciton–Vibration and Exciton–Photon Interactions in Organic Semiconductors*, Springer Theses,
https://doi.org/10.1007/978-3-030-85454-6_7

tems [2, 3] to organic solar cells [4]. Following photoexcitation, ultrafast (i.e. femto- to picosecond timescale) non-equilibrium dynamics drive the exciton energy transfer in these systems [5], underlining the need to move away from the traditional thermodynamic way of thinking, in order to better understand and engineer such processes. In the past decade, the experimental study of organic semiconductors has provided a good understanding of ultrafast processes, through the use of techniques such as pump-push probe spectroscopy [6–8] and transient absorption microscopy [9, 10]. Several studies have emphasised the role of strong exciton-phonon coupling for driving energy transfer in these systems [11–14], indicating the relevance of non-Markovian physics for such phenomena.

Following photoexcitation, one may think of the interaction of an exciton system with the potentially large number of molecular vibrations of organic structures as the interaction of an open quantum system with its (vibrational) environment. The various experimental setups outlined previously, allow for preparing different initial quantum states, the time evolution of which can be followed with great precision in the fs timescale. Examples include 'hot' states, i.e. exciton states with excess energy pumped into molecular vibrations [15, 16]. The insights generated from such works may be generalised to a wide range of open quantum systems, which are relevant for quantum technologies, biology and areas such as nanoscale quantum heat engines [17–22]. Furthermore, by formulating the problem within the framework of open quantum system theory, several techniques from these fields of study become available in order to study the non-equilibrium dynamics from a theoretical perspective. Such open quantum system treatments of organic molecules have been used in the past [23, 24], including techniques ranging from reduced density matrix descriptions (master equations) [25–27] to complete simulations of system-bath wave functions with methods such as multi-configurational time-dependent Hartree (MCDTH) [28–30].

However, the theoretical and computational study of these systems poses a significant challenge for several reasons. Firstly, the large number of molecular vibrations often found in organic molecules makes it necessary to account for complicated many-body interactions. The strong exciton-phonon coupling which was mentioned previously further complicates things, as traditionally used perturbative approaches are not necessarily applicable in this case [14, 31]. For these reasons, we adopt a tensor network formulation [32] based on the formalism of matrix product states [33], which has previously been used to study the ultrafast dynamics of singlet exciton fission [34] and exciton-polariton dynamics [35, 36]. This so-called tree tensor network state (TTNS) approach shows strong similarities to the multilayer formulation of MCTDH [37], and can capture the full non-Markovian quantum dynamics of a system coupled to a large number of vibrations (few hundreds), without relying on any perturbative approach. Moreover, it efficiently retains all information about the vibrational modes of the environment, enabling us to identify and visualise the often complex chain of environmental processes that drive ultrafast dissipative dynamics.

Here we use our TTNS formalism to explore the dynamics of systems prepared far from equilibrium, such that excess vibrational energy could potentially promote transient excitonic dynamics that lead to new outcomes, which would not occur for band-edge excitation (i.e. no excess vibrational energy). The role of such 'hot' states has been a subject of debate in the literature, in particular regarding the process of charge transfer and separation, in materials used for organic photovoltaics [15, 16, 38, 39]. However, their importance extends to several temperature-activated processes, such as endothermic singlet exciton fission [40] and energy transfer in biological systems [41, 42]. Our results indicate that excess energy can have a large impact on the early timescales of non-equilibrium dynamics, leading to very different yields for the process of charge transfer in a molecular dimer of the organic semiconductor tetracene which we focus on in this work. Perhaps most intriguingly, we find that initial quantum states where low-frequency modes are excited preferentially lead to greater charge transfer yields, due to robust, mode-generated quantum coherence between the initial exciton and the charge transfer state. Indeed, modes of different frequencies have been discussed to play different roles in the coherent dynamics of charge transfer [43], while the electron mobility in a Holstein model has been shown to strongly depend on the initial vibrational preparation [44]. We relate the behaviour observed in our system to the different pathways of vibrational relaxation which the system follows, depending on the energy of the excited vibrations. Overall, we believe that our study suggests a more general principle for engineering the dynamics of open quantum systems, and helps open a way to understanding how excess energy, heat generation, work can be managed in quantum optoelectronic devices. Encouragingly, recent experimental advances in the ultrafast preparation and control of particular vibrational modes in excited states, including the exploitation of ultra-strong light-vibration coupling (polaritons), suggest that the underlying microscopic physics now available in simulations of quantum internal conversion may be tested in the relatively near-term [45, 46].

The structure of this chapter is as follows. In Sect. 7.2 we outline the used methods. Section 7.2.1 provides a general introduction to tensor network methods for studying the dynamics of open quantum systems, as they have been developed both in previous and in this work. Section 7.2.2 deals with the specifics of applying this formalism to a molecular system, for which it is necessary to obtain its electronic and vibrational properties. The results on the ultrafast dynamics of charge transfer in this molecular system are presented in Sect. 7.3, comparing the cases of having finite or zero excess energy in the system, under different conditions. The conclusions of the chapter are finally given in Sect. 7.4.

7.2 Methodology

7.2.1 Tensor Networks

Matrix product states

In order to develop some intuition for the tree-tensor network state (TTNS) ansatz that we shall use to simulate molecular open system dynamics, we will begin with a simpler illustration using matrix product states (MPS). A comprehensive review of the properties and uses of MPS, as well as their fundamental links to Density Matrix and Wilsonian Renormalisation Group algorithms, can be found in Ref. [33]. Let us consider a one dimensional many-body system of L (distinguishable) interacting particles, each of which is localised on a lattice 'site' k and has a local (non-interacting) Hilbert space basis $\{|n_k\rangle\}$ of dimension d_k. The exact many-body wave function may be formally written in the Kronecker product basis as:

$$|\Psi\rangle = \sum_{\{n_k\}=1}^{d_k} \Psi_{n_1,\ldots,n_L} |n_1, \ldots, n_L\rangle , \tag{7.1}$$

where $|n_1, \ldots, n_L\rangle$ is shorthand for the tensor product $|n_1\rangle \otimes |n_2\rangle \ldots \otimes |n_L\rangle$, and the sum runs over all possible products of n_k. The multidimensional array, here referred to as a tensor, $\Psi_{n_1,\ldots,n_L}$ contains the probability amplitudes for all possible states of the chain. As the number of particles grows, the number of elements stored in this tensor scales as d_k^L, and will quickly become too large for practical computations. This is often called the curse of dimensionality. However, by iteratively applying singular value decomposition (SVD), any $|\Psi\rangle$ may be written as a product of rank three tensors, known as a Matrix Product State (MPS) [33]:

$$|\Psi_{MPS}\rangle = \sum_{\{n_k\}=1}^{d_k} A^{n_1} \cdot \ldots \cdot A^{n_L} |n_1, \ldots, n_L\rangle . \tag{7.2}$$

Each of these tensors A^{n_k} has dimensions D_{k-1} x D_k x d_k, with D_{k-1}, D_k referred to as the left and right bond dimensions, and d_k the bond dimension of the local Hilbert space. The maximum value of D_k encodes the amount of entanglement between neighbouring sites. Since for every configuration the tensors must multiply into a scalar, we have the boundary condition: $D_1 = D_L = 1$. The procedure of writing a many-body wavefunction as an MPS is summarised in Fig. 7.1. The local Hilbert space of each site may be compressed by using an optimised boson basis (OBB) [47, 48] with $d_{OBB,k} << d_k$:

$$A^{n_k} = \sum_{\tilde{n}_k}^{d_{OBB,k}} A^{\tilde{n}_k} V_{n_k,\tilde{n}_k} \tag{7.3}$$

Once a state has been written as an MPS, any variational optimisation or time-evolution may be performed by 'sweeping along the chain', i.e. updating one tensor at a time, in a manner similar to density matrix renormalisation group (DMRG) approaches [33]. The MPS approach thus benefits from the reduction of the problem to a sequence of single site problems and the storage and manipulation of a number of matrices that scales linearly with system size. When we account for the overheads due to the contraction of tensors, single value decomposition, etc., the effective scaling becomes cubic in system size, but this is still a vast improvement on the previous

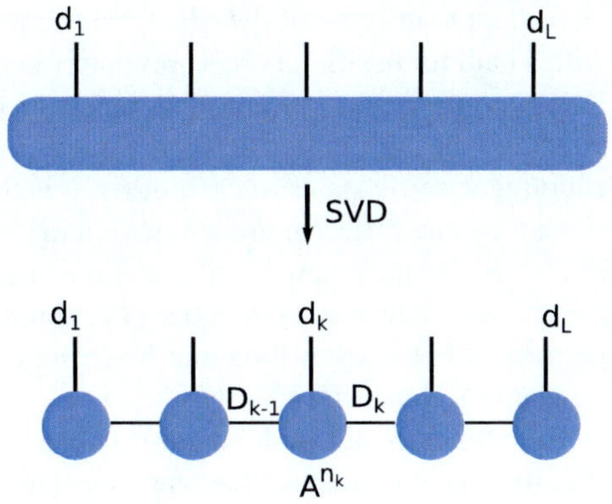

Fig. 7.1 A matrix product state (MPS) is constructed from a many-body wavefunction by applying singular value decomposition (SVD) iteratively

exponential scaling. Discussions and bench marking data related to system sizes, convergence, and CPU times for spin-boson-like systems can be founds in Refs. [34, 48].

Systems interacting with a vibrational environment

Let us examine the case of a system which is in contact with a vibrational 'bath', i.e. a set of oscillators with which it can exchange energy. The Hamiltonian of such a system may be schematically represented as in Fig. 7.2 (left). Such a system does not have a one-dimensional architecture which would allow a one-to-one match between the different components of the Hamiltonian and the MPS tensors. However, even for such a system, the problem may be recast in that of a one-dimensional system using an orthogonal polynomial mapping [49]. Such an approach has previously been employed to study the well known Spin Boson Model (SBM) in the case where it is coupled to an environment of oscillators [48, 50, 51]. This process is visualised in Fig. 7.2. The bond dimension between the system tensor and the environment chain (D_h), may in principle be different from the intra-chain bond dimension (D).

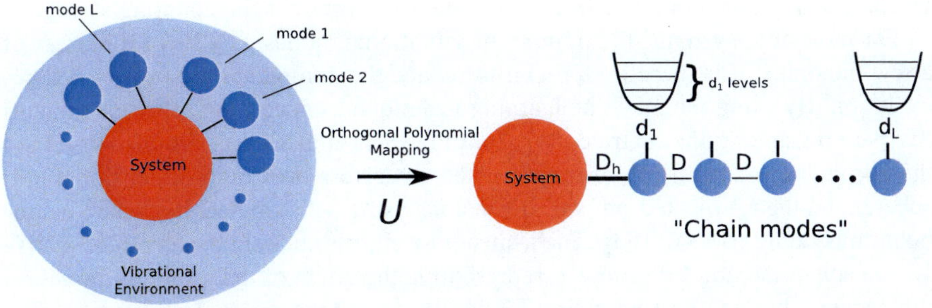

Fig. 7.2 A system interacting with a set of oscillators (left) may be transformed into a 'chain architecture' (right) using an orthogonal polynomial mapping U. The new modes are linear combinations of the initial ones, and we refer to them as 'chain modes'

Similar to the case of the SBM, this approach of representing a system interacting with a bath for oscillators has previously been employed for molecular systems [34]. However, in such a system, different molecular vibrations can have very different effects on the system states, e.g. only shifting the energies of particular levels, or only coupling specific states to each other. It is therefore necessary to separate molecular vibrations into different groups, which in [34] approximately correspond to different irreducible representations of the molecular point group. However, when there are no obvious molecular symmetries to exploit, this grouping of vibrations may still be performed in a rigorous manner by using the machine learning technique known as k-means++ clustering [52].

In order to better understand this procedure, let us work with a linear vibronic Hamiltonian which describes the coupling of the system states to the vibrations:

$$H^{m,n} = H_{el}^{m,n} + \sum_k W_k^{m,n} Q_k + \sum_k \frac{\hbar\omega_k}{2}\left(-\frac{\partial^2}{\partial Q_k^2} + Q_k^2\right), \tag{7.4}$$

where Q_k are the dimensionless displacements along normal mode k, indicating the contribution of that mode to the molecular deformation. The matrices W_k give the first order couplings between the system states due to deformation along mode k, and will hence be referred to as vibronic coupling matrices. In terms of creation and annihilation operators:

$$H = H_{el} + \sum_k \left(W_k \frac{a_k^\dagger + a_k}{\sqrt{2}} + \hbar\omega_k a_k^\dagger a_k \right). \tag{7.5}$$

By normalising the coupling matrices W_k as:

$$(W_{mn})_k = (\hat{W}_{mn})_k \lambda_k, \tag{7.6}$$

the matrix $(\hat{W}_{mn})_k$ contains the coupling pattern between the system states due to displacement along mode k, while the constant λ_k describes the coupling strength.

For molecular systems, the number of vibrational modes may be of the order of a few hundreds, it is therefore crucial to reduce the computational cost of a time-evolution. By using the machine learning technique k-means++ [52], we can identify patterns among the matrices W_k, splitting them into several groups. To do so, the normalised matrices $\hat{W}_k$ are represented as vectors on a high-dimensional unit sphere, and then projected on two dimensions using t-distributed stochastic neighbour embedding (t-SNE) [53]. The k-means++ algorithm assigns a centroid matrix $\bar{W}_i$ to each of the created groups i, representing the effect of the group of modes on the system. This is visualised in Fig. 7.3, for the case of the molecular system studied in this work (see below). The individual modes still retain their different coupling strengths λ_k, but now act as one 'environment', which can be transformed into a one-dimensional chain of oscillators using an orthogonal polynomial mapping. Hence the linear vibronic Hamiltonian may be written in its final form:

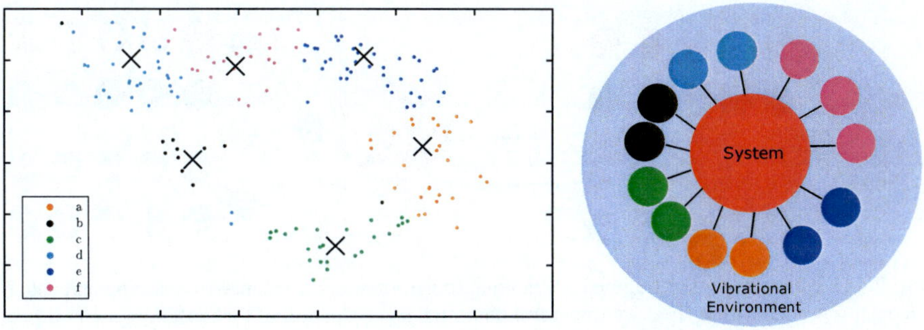

Fig. 7.3 Clustering of molecular vibrations using k-means++ [52]. We find that a minimum of six clusters is required to obtain converged dynamics for our system of study. The modes belonging to the different clusters are represented using different colours

$$H_{star} = H_{el} + \sum_i \bar{W}_i \sum_k \lambda_{i,k} \frac{b^\dagger_{i,k} + b_{i,k}}{\sqrt{2}} + H_{c,i}, \tag{7.7}$$

where $H_{c,i}$ is the Hamiltonian of the transformed vibrational modes of chain i. The creation and annihilation operators $b^\dagger_k$, b_k now refer to the 'chain modes' which result from the orthogonal polynomial transformation of the original system vibrations.

Time evolution and tree tensor network states

Once we apply an orthogonal polynomial mapping to every vibrational environment of the linear vibronic Hamiltonian and bring it in the form of Eq. (7.7), it may be schematically represented as in the left hand side of Fig. 7.4. We represent the many-body wavefunction as a tensor network which resembles this same architecture; this is essentially several MPS connected to each other through the central system (red) tensor. We refer to this wavefunction as the 'star MPS'.

The time-evolution of a wavefunction written as a star MPS in Fig. 7.4 (left), may be done using the time-dependent variational principle [34, 54]. In short, each tensor is sequentially time-evolved using a local effective Hamiltonian, which is constructed by contracting the full many-body Hamiltonian with all MPS tensors which are not currently being updated. Unfortunately, this star MPS suffers a curse of dimensionality: due to being connected to each other through the central tensor, all possible environment configurations are entangled, making its size to scale exponentially with the number of environment chains.

In order to address the exponential scaling problem, the star MPS tensor might be further decomposed into a number of smaller, entanglement renormalisation tensors (ER nodes), which describe the inter-environment entanglement [55]. The entanglement of the various bipartitions of the system tensor is quantified by calculating the corresponding von Neumann entropy. The tensors are then connected into a tree structure, the tree tensor network state (TTNS), according to the coupling structure of the star Hamiltonian, in order to minimise the total von Neumann entropy. To

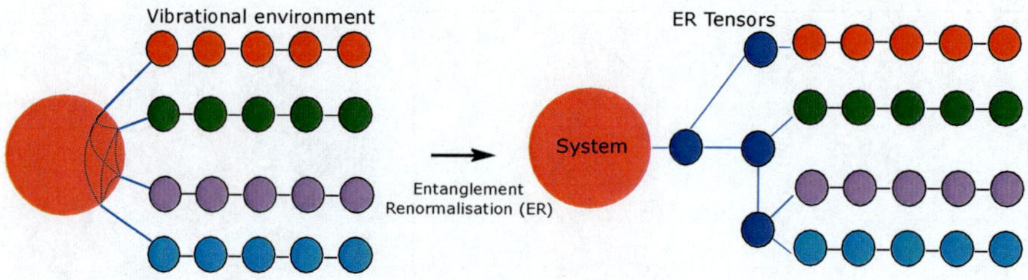

Fig. 7.4 By means of an orthogonal polynomial transformation, k-means++ clustering and entanglement renormalisation (ER) we transform the initial wavefunction of the system in a vibrational bath, into a tree tensor network state (TTNS)

find the optimal tree structure, an initial simulation with the star MPS is performed, allowing us to assess the entropy of various possible tree structures, and construct the one with the minimal value. The TTNS obtained through this analysis allows for using the smallest possible values for bond dimensions, in order to capture the entanglement of the system. The size of this tensor network now scales linearly, which is a significant improvement compared to the initial exponential scaling. The process for determining the TTNS is summarised in Fig. 7.4.

At this stage, it is important to remember that the time-evolution is performed on the system after the application of an orthogonal polynomial mapping was applied to each of the vibrational environments. Therefore, in order to obtain the time-dependent observables of the original system, an inverse mapping is required in each case.

Constructing initial states

Finally, we would like to discuss a few ways of constructing an initial TTNS, which correspond to various different physical initial conditions. Since in this work we are concerned with vibrational relaxation and the effects of adding excess energy in the system, i.e. initially exciting not only the 'system' but also certain vibrations of the 'bath', let us consider a few possible ways of defining a vibrationally excited TTNS as our initial state.

1. **Band-edge excitation.** By band-edge excitation, we mean an initial state where a system state is excited, however no excess energy is added to the vibrational bath. By denoting each tensor within our MPS representation as $A(D_{k-1}, D_k, d_k)$, we set $A(1, 1, 1) = 1$ and all other values to zero. Therefore, from the sum of Eq. (7.2), only the term with all vibrational modes in their ground state, and the system in the first of the two states, survives. We remind at this point that d_k denotes the bond dimension of the local Hilbert space of site k, while the bond dimensions D_{k-1}, D_k encode the entanglement which the MPS can describe between neighbouring sites. For tensors representing vibrations, we find that a maximum value of $d_k = 40$ leads to converged dynamics for all cases examined in this chapter. This means that we allow each site to accommodate up to 40 Fock states. For the special case of the two-state exciton system $d = 2$. We also find that for the intra-chain and system-environment entanglement (see Fig. 7.2), the

bond dimensions $D = 35$ and $D_h = 50$ produce converged results respectively (refer to Appendix 1 for convergence tests).

2. **Exciting a 'chain mode'**. The most straightforward way of adding excess energy to the system is by adding a quantum of energy to one of the tensors l representing vibrations on the right hand side of Fig. 7.4: $A_l(1, 1, 1) = 0$ and $A_l(1, 1, 2) = 1$, i.e. we populate the first excited state (second Fock state, $d_k = 2$) of the mode at site l. However, one has to bear in mind that these are not the original, physical vibrations of the system, but the ones resulting from the orthogonal polynomial transformation. The energy corresponding to one quantum of such a chain mode may be derived analytically, as done in [48]. Since the excitation of these 'chain modes' does not correspond to physical reality, we do not pursue this avenue, but rather develop two different methods for adding energy to the original molecular vibrations.
3. **Incoherent (quantised) excitation of a normal mode**. Let us start by describing the example case of defining an initial MPS, which describes a state with one quantum of excitation added to mode k of the original system. The orthogonal polynomial transformation discussed previously, may be used to write the annihilation (or creation) operator of 'chain mode' i, in terms of the annihilation operators of the physical normal modes:

$$b_i = \sum_k U_{ik} a_k, \tag{7.8}$$

Since the mapping U is unitary, it is straightforward to obtain the operators of the original modes in terms of chain modes:

$$a_k = \sum_i U^*_{ki} b_i. \tag{7.9}$$

Therefore, the state with one quantum on mode k is a linear combination of states with one quantum on the various chain modes i, and coefficients given by the inverse of the orthogonal polynomial mapping ($U^*_{ki} = U_{ki}$):

$$|1\rangle_k = a^\dagger_k |0\rangle = \sum_i U_{ki} b^\dagger_i |0\rangle . \tag{7.10}$$

To write this in the form of an MPS, one has to realise a way in which the sum over all possible configurations of Eq. (7.2) produces exactly the state of Eq. (7.10). All the configurations describing states which do not contribute to this sum, for example states with two quanta on different modes, should vanish. To achieve this for the case of one quantum, one may define the tensors within the chain as:

$$A_i(:, :, 1) = \begin{bmatrix} 1 & 0 \\ 0 & 1 \end{bmatrix}, A_i(:, :, 2) = \begin{bmatrix} 0 & U_{ik} \\ 0 & 0 \end{bmatrix} \tag{7.11}$$

Since matrices need to multiply into a scalar for each configuration, the last site L of a chain is always represented by column vectors:

$$A_L(:,1) = \begin{bmatrix} 0 \\ 1 \end{bmatrix}, A_L(:,2) = \begin{bmatrix} U_{ik} \\ 0 \end{bmatrix}. \tag{7.12}$$

The first site of the chain is the tensor representing the system, which is correspondingly represented as a row vector of dimensions 1×2 in this case. In this manner, if site i is in its excited state represented by the second matrix of Eq. (7.11), then if any other site is in its excited state the probability of encountering this state is zero, since the product of two such matrices is zero. Only states with a single excited chain mode multiplying chain modes in their ground state will survive in the sum of Eq. (7.2), and will have a prefactor of U_{ik}. Therefore we obtain exactly the state of Eq. (7.10). By considering the last site of the chain L, this becomes rather easy to see, by asking the question of which 2×2 matrix needs to operate on each column vector in order for the product to vanish/survive. Before generalising to N quanta excitation, let us consider one more example case of adding two quanta on mode k:

$$|2\rangle_k = \frac{1}{\sqrt{2}}(a_k^\dagger)^2 |0\rangle = \frac{1}{\sqrt{2}}\left(\sum_i U_{ki} b_i^\dagger\right)^2 |0\rangle . \tag{7.13}$$

Other than the possibility of having two quanta on chain mode i, we now also need to allow for two different chain modes i and j accommodating a quantum each, as described by the cross terms which appear once we take the square of the sum of Eq. (7.13). It is easier to start from the MPS tensors of the last chain site:

$$A_L(:,:,1) = \begin{bmatrix} 0 \\ 0 \\ 1 \end{bmatrix}, A_L(:,:,2) = \begin{bmatrix} 0 \\ U_{ik} \\ 0 \end{bmatrix}, A_L(:,:,3) = \begin{bmatrix} \frac{U_{ik}^2}{\sqrt{2}} \\ 0 \\ 0 \end{bmatrix}, \tag{7.14}$$

which allow for the possibility of having two, one or zero quanta on the last chain mode of a chain. The vector $A_L(:,:,3)$ needs to only give non-zero entries once it encounters matrices of other sites in their ground state, $A_L(:,:,2)$ in the case where the other sites have one or zero quanta, and $A_L(:,:,1)$ in all cases. Therefore, for the intermediate sites i of the chain:

$$A_i(:,:,1) = \begin{bmatrix} 1 & 0 & 0 \\ 0 & 1 & 0 \\ 0 & 0 & 1 \end{bmatrix}, A_i(:,:,2) = \begin{bmatrix} 0 & U_{ik} & 0 \\ 0 & 0 & U_{ik} \\ 0 & 0 & 0 \end{bmatrix}, \tag{7.15}$$

$$A_i(:,:,3) = \begin{bmatrix} 0 & 0 & \frac{U_{ik}^2}{\sqrt{2}} \\ 0 & 0 & 0 \\ 0 & 0 & 0 \end{bmatrix}.$$

From the way we defined the initial MPS for the cases of one and two quanta, the reader may already start to identify a pattern going towards higher excitations. We generalise for the case of N quanta excitation by defining the non-zero entries of the initial MPS as:

$$A_L(N-m+2,1,m) = \frac{U_{ik}^{m-1}}{\sqrt{(m-1)!}}, \tag{7.16}$$

where m runs over all integers from 1 to $N+1$. For intermediate chain sites i:

$$A_i(1,N+1,N+1) = \frac{U_{ik}^{m-1}}{\sqrt{(m-1)!}}, A_i(l,l,1) = 1, \tag{7.17}$$

where l obtains integer values from 1 to $N+1$. Then:

$$A_i(N,N+1,m) = \frac{U_{ik}^{m-1}}{\sqrt{(m-1)!}}, A_i(1,m,m) = \frac{U_{ik}^{m-1}}{\sqrt{(m-1)!}} \tag{7.18}$$

and finally:

$$A_i(1+j,m+j,m) = \frac{U_{ik}^{m-1}}{\sqrt{(m-1)!}}, \tag{7.19}$$

with j obtaining all integer values from 1 to $N-m$. As opposed to the case of exciting a coherent state, which is a linear combination of different Fock states and is described in the following subsection, the process outlined here only results in the excitation of one Fock state. To differentiate between the two, we refer to this case as 'incoherent' excitation, as opposed to the coherent excitation which we now proceed to discuss.

4. **Coherent (continuous) excitation of a normal mode**. Rather than adding a set number of quanta to a selected normal mode, one can displace a vibration continuously, by setting its initial dimensionless displacement Q appearing in the linear vibronic Hamiltonian of Eq. (7.4) to a finite value $Q=\alpha$. However, how to do this when working in the chain representation might not be immediately obvious.

In order to create a displaced state $|\alpha\rangle$ from the vacuum state $|0\rangle$ one may apply the displacement operator:

$$D(\alpha) = \exp(\alpha a^\dagger - \alpha^* a), \tag{7.20}$$

as $|\alpha\rangle = D(\alpha)\,|0\rangle$. The state $|\alpha\rangle$ is commonly referred to as a coherent state. The expectation values of its position and momentum oscillate just like those of a classical harmonic oscillator initially displaced by α, and it corresponds to the state excited by a coherent laser pulse [56].

One may use the inverse orthogonal polynomial mapping of Eq. (7.9), and write the above displacement operator for mode k in terms of the chain modes:

$$D_k = \prod_i \exp(\alpha \cdot U_{ki}(b_i^\dagger - b_i)). \tag{7.21}$$

Therefore, we only have to apply this operator to the MPS defined for the case of band-edge excitation above, in order to obtain an initial state where mode k is displaced by α. Each of the exponential operators appearing in the product of Eq. (7.21) may be written as a $d_i \times d_i$ square matrix. By writing the coherent state in the basis of Fock states:

$$|\alpha\rangle = e^{-\frac{|\alpha|^2}{2}} \sum_n \frac{\alpha^n}{\sqrt{n!}} |n\rangle \tag{7.22}$$

one realises that for larger n values, the terms of the sum become increasingly small. It is therefore possible to truncate a number of terms, corresponding to reducing the second dimension of the matrix representation of the exponential operators of Eq. (7.21). This is similar to the transformation of the Fock basis into the optimised boson basis through the matrices V appearing in Eq. (7.3), therefore we denote these truncated exponential matrices as having dimensions $d \times d_{OBB}$. We find that choosing $d_{OBB} = 20$ allows for correctly describing initial coherent states with displacements of up to $\alpha = 6$, while still maintaining a reasonable computational cost.

7.2.2 *Studied System*

We study a covalent dimer of the organic semiconducting molecule tetracene, shown in Fig. 7.5. Tetracene is a prototypical organic semiconductor, best known for its ability to efficiently undergo an ultrafast process known as singlet fission [57, 58], despite the process being endothermic [59, 60]. For covalently linked systems as the one at hand, it is common practice to introduce Mesityl side groups for solubility reasons [61]. We therefore refer to this di-tetracene-Mesityl system as DT-Mes in short.

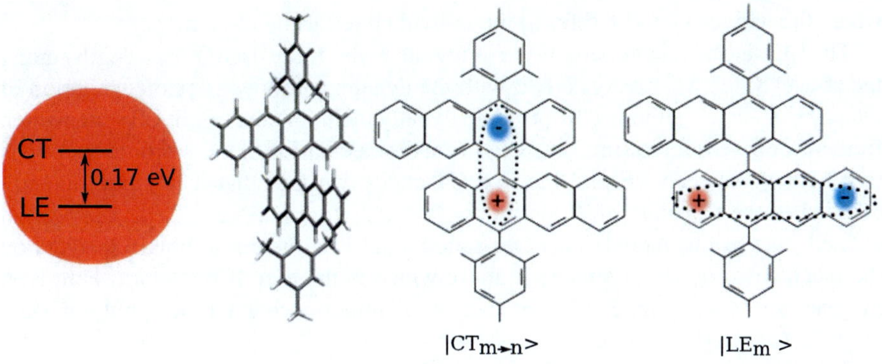

Fig. 7.5 Studied system includes an electronic system of two excitons, a local exciton (LE) and a charge transfer exciton (CT), coupled to an environment of 156 oscillators. The oscillators correspond to molecular normal modes. The indices *m* and *n* refer to the individual tetracene monomers

We optimise the geometry of DT-Mes using density functional theory at the B3LYP, cc-pVDZ level. We find that the molecule assumes an orthogonal configuration due to the steric repulsion between the hydrogen atoms on the two tetracene monomers. The molecular vibrations are obtained at the same level of theory. Concerning the excited states of this system, we perform PPP calculations [62, 63] at the ground state geometry, which have successfully been used in the past to describe conjugated and singlet fission systems [57, 64, 65]. Because the Mesityl side groups break the conjugation of the molecular structures, we truncate them in order to be able to apply PPP theory. However, both excited states considered in this study are localised on the tetracene monomers (see below), so this truncation has a negligible effect on the excited state energies.

Due to the approximate C_2 symmetry of the molecule, we find that the excited states transform either as B or A irreducible representations of the point group. There are two kinds of excitations in such a system, visualised in Fig. 7.5: a local exciton (LE), i.e. a state where a bound electron-hole pair is localised on a monomer, and a charge transfer (CT) exciton, where the electron and hole are localised on different monomers. Charge transfer excitons are particularly relevant for light-harvesting applications, as they are the precursor to a charge separated state. In terms of the excitons visualised in Fig. 7.5, the symmetry adapted states may be written as:

$$
\begin{aligned}
|\mathrm{LE}_A\rangle &= \frac{1}{\sqrt{2}}(|\mathrm{LE}_m\rangle + |\mathrm{LE}_n\rangle) && (A \text{ symmetry}) \\
|\mathrm{LE}_B\rangle &= \frac{1}{\sqrt{2}}(|\mathrm{LE}_m\rangle - |\mathrm{LE}_n\rangle) && (B \text{ symmetry}) \\
|\mathrm{CT}_A\rangle &= \frac{1}{\sqrt{2}}(|\mathrm{CT}_{m\to n}\rangle + |\mathrm{CT}_{n\to m}\rangle) && (A \text{ symmetry}) \\
|\mathrm{CT}_B\rangle &= \frac{1}{\sqrt{2}}(|\mathrm{CT}_{m\to n}\rangle - |\mathrm{CT}_{m\to n}\rangle) && (B \text{ symmetry}),
\end{aligned}
\tag{7.23}
$$

where the indices m and n refer to the individual tetracene monomers.

The LE state of B symmetry is the only state with a finite oscillator strength among the above. Since we are concerned with the dynamics following photoexcitation of the bright state, we simplify the problem by only including the bright LE state, which from now on we refer to simply as LE. From the calculation of the vibronic couplings (see below) between LE and the two CT states, we find that it is predominantly coupled to the CT state of B symmetry. We therefore choose CT_B as the second excited state of our model. This two-state model is sufficient to obtain insights on the mechanism of vibrational relaxation, which is the aim of this work. However, in general, the A symmetry states can be of importance for other photophysical processes as singlet fission.

The CT state lies 0.17 eV above the LE state, hence charge transfer from the bright exciton is an endothermic process. At the ground state geometry, there is a small but finite electronic coupling between the LE and CT states, and in the basis $\{|\mathrm{LE}\rangle, |\mathrm{CT}\rangle\}$, the electronic Hamiltonian may be written as:

$$H_{el} = \begin{bmatrix} 2.39 & -0.014 \\ -0.014 & 2.56 \end{bmatrix} \text{ (eV)}. \tag{7.24}$$

The fact that the electronic Hamiltonian is not diagonal means that the adiabatic states of the system are linear combinations of the LE and CT states. However, since this coupling is small, the first and second excited states are of almost pure LE and CT character respectively. In this basis, the reduced density matrix of the system may be written as:

$$\rho_{nm} = \begin{bmatrix} \rho_{\mathrm{LE,LE}} & \rho_{\mathrm{LE,CT}} \\ \rho_{\mathrm{LE,CT}} & \rho_{\mathrm{CT,CT}} \end{bmatrix}, \tag{7.25}$$

where the diagonal entries give the probability of finding the system in the LE or CT state, while the off-diagonal element is the quantum coherence between the two excitons.

The two excitons are coupled to a bath of oscillators, which are the molecular normal modes. Out of the 210 vibrational modes of the truncated DT-Mes structure, we include a total of 156 in our model. These are modes with frequencies between 100 and 1600 cm^{-1}, i.e. we exclude very high-frequency modes, mostly C-H stretches, which are not relevant for the dynamics, as well as very slow low-frequency modes, which are not strongly displaced in the ultrafast timescale. In addition, low-frequency modes are known to be strongly anharmonic [66], and cannot be treated within the harmonic approximation, on which the linear vibronic model relies.

In order to obtain the coupling between the electronic states upon displacement of the vibrational modes, i.e. the vibronic couplings, we displace the molecular structure in the positive and negative direction of each normal mode $\mathbf{q}$, and recalculate the electronic states using PPP calculations. The coupling element between two states is then given by:

$$W_{mn} = \frac{\langle m|\,\mathcal{H}(\mathbf{r}_o + h\mathbf{q})\,|n\rangle - \langle m|\,\mathcal{H}(\mathbf{r}_o - h\mathbf{q})\,|n\rangle}{2h}. \tag{7.26}$$

We thus construct a matrix W for each of the 156 vibrations included in the model, and apply our previously described clustering algorithm on them. We find that in order to obtain converged results, a minimum of six clusters is required, schematically depicted in Fig. 7.3.

The above concludes the necessary steps for parametrising the linear vibronic Hamiltonian of Eq. (7.7) for DT-Mes. We now proceed to use this Hamiltonian for a time evolution of the system.

7.3 Results

7.3.1 Band-Edge Excitation

We start by discussing the population dynamics upon band-edge excitation of the two excitons, i.e. with the system excited and the vibrational modes in their ground state. Figure 7.6 visualises these population dynamics. For excitation of the LE state, we barely see any CT formation. On the other hand, exciting the higher-energy CT state leads to relaxation towards LE, within a timescale of approximately 1 ps.

In molecular systems such as the one studied here, the CT state is typically dark [61], so exciting it directly is not possible. Instead, one has to excite a 'hot' LE configuration, i.e. excite LE beyond its band-edge, giving excess energy to molecular vibrations. This can open a channel to CT formation. To identify specific vibrations which couple the LE and CT excitons, we plot in Fig. 7.7 the amplitude of modes displaced by $Q = 0.2$ or more over the course of the relaxation from CT to LE. We find that a low-frequency mode, with a frequency of $\omega = 276\,\text{cm}^{-1}$, is the most active motion. Regarding the real-space motion of this mode, it involves a strong oscillation of the central bond connecting the two tetracene monomers. The strong displacement of this mode upon the CT to LE conversion makes it a good candidate to excite, in

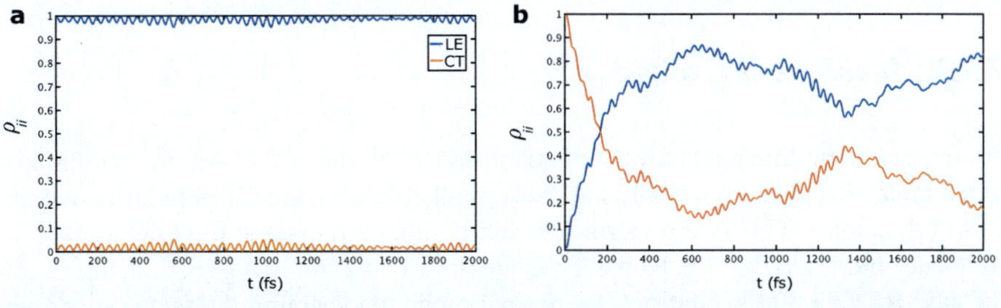

Fig. 7.6 Population of the electronic states in real-time, in the cases of an initially excited LE (panel **a**) and CT (panel **b**) state

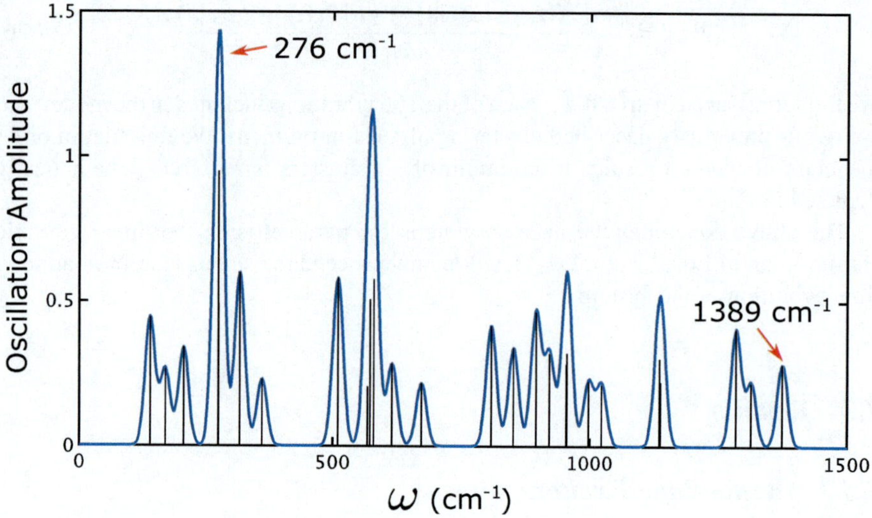

Fig. 7.7 Amplitude of the most displaced molecular vibrations, during 2 ps of dynamics following CT excitation

order to study the effects of excess energy on CT formation. From now on, we refer to this vibration simply as 'the low-frequency mode'. Its coupling constant appearing in the linear vibronic Hamiltonian of Eq. (7.7) is equal to $\lambda = -289\,\text{cm}^{-1}$.

Furthermore, in order to compare the effects of adding excess energy to the system through a high- and low-frequency mode, we choose the vibration at $\omega = 1389\,\text{cm}^{-1}$ as another candidate for excess energy excitation. We refer to this mode as 'the high-frequency mode'. This is a breathing motion of the tetracene carbon rings. Its coupling constant is $\lambda = 256\,\text{cm}^{-1}$, a value not very far from the one of the low-frequency mode introduced above. We can hence be confident that any differences between the cases of exciting the low- and high-frequency modes are not due to differences in the magnitude of the exciton-phonon coupling constants. The two modes also belong to the same group of modes from the clustering algorithm, so the pattern of their coupling expressed through the matrix W of Eq. (7.7) is identical.

7.3.2 *Incoherent Excitation*

For the case of adding a set number of quanta to a Fock state of the original vibrations of the molecular system, we only find very small effects on the CT population within 2 ps of dynamics. This is shown for the case of adding quanta to the low- and high-frequency modes in Fig. 7.8. However, as we show in the following section, this is not the case for coherent excitation of a normal mode, underlining the fact that adding excess energy to a system is not enough by itself, but it needs to be done in the right way in order to facilitate processes of interest.

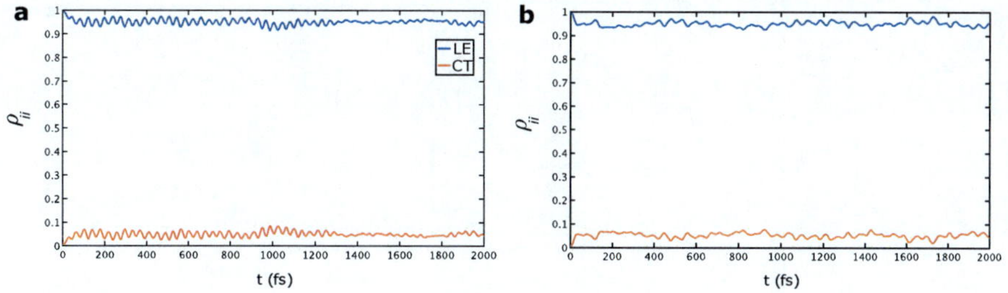

Fig. 7.8 Population dynamics upon LE excitation with excess energy added in the system incoherently, through **a** three quanta of a high-frequency mode ($\approx 0.51\,\text{eV}$) and **b** seven quanta of a low-frequency mode ($\approx 0.24\,\text{eV}$)

7.3.3 Coherent Excitation

For the case of exciting the two modes coherently as presented in the methodology Sect. 7.2, we start by visualising the population of the two exciton states. Figure 7.9 shows the resulting dynamics, once the high- and low-frequency modes are displaced by $Q = 4$. From the linear vibronic Hamiltonian of Eq. (7.7), it becomes apparent that for the same displacement, the high-frequency mode carries significantly more energy than the low-frequency vibration. For $Q = 4$, the two modes carry 1.36 eV and 0.27 eV of energy respectively.

Comparing the two graphs of Fig. 7.9, we see that in both cases excess energy opens a channel to CT formation, unlike the case of incoherent excitation summarised in Fig. 7.8. It therefore becomes clear that in order for excess energy excitation to have an effect on the population dynamics of the two-level system, one needs to induce a coherent displacement of vibrational modes, i.e. excite a wavepacket with finite displacement Q. The dynamics towards this CT formation can take longer than the 2 ps window that we have examined thus far, we therefore simulated the system for a total of 4.5 ps in the cases of coherent excitation.

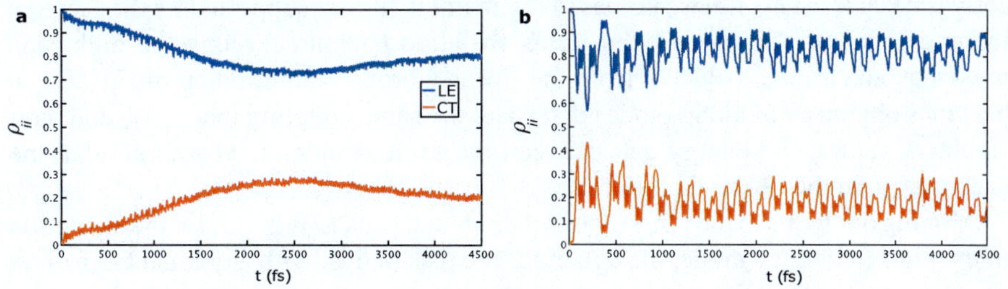

Fig. 7.9 Population dynamics upon LE excitation with excess energy added in the system coherently, through **a** a high- and **b** a low-frequency mode vibration. A dimensionless displacement $Q = 4$ was chosen for both vibrations, corresponding to excess energy of 1.36 eV and 0.27 eV respectively

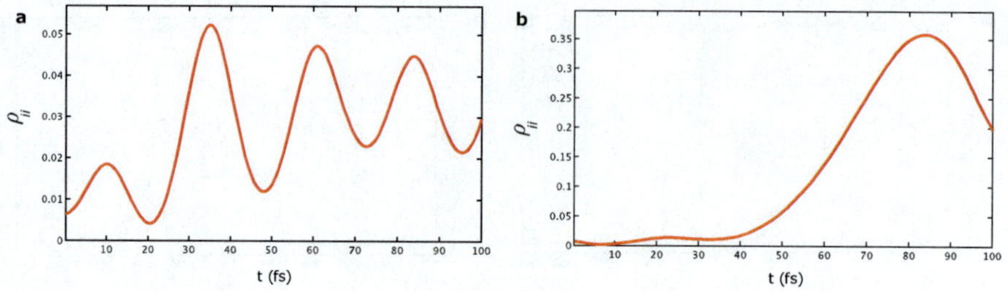

Fig. 7.10 Population of the CT state at early times, once excess energy is added in **a** a high-frequency and **b** a low-frequency mode

Furthermore, the two cases of coherent excitation in Fig. 7.9 exhibit some striking differences, both at early and at later times. Let us start by discussing the early-time (< 100 fs) differences between the two cases. Figure 7.10 provides a close view of the CT population during these first 100 fs of dynamics. We observe that exciting the low-frequency mode leads to a stronger increase of the CT population (approximately 35% of CT at 90 fs), in what seems to be an almost steady population growth. Instead, in the high-frequency case, the CT population is oscillatory and does not go beyond 5%. We believe this difference to be due to the time which is required for the crossing event from the LE to the CT surface to occur. The rate of the LE to CT population transfer is dictated by the off-diagonal entries of the linear vibronic Hamiltonian of Eq. (7.7). All vibrational modes contribute to this value through the linear term, however it is the electronic coupling of 14 meV which appears in the electronic part of the Hamiltonian (see Eq. 7.24) that dominates. This electronic coupling dictates a timescale of approximately 300 fs for the crossing event from the LE to the CT surface, a timescale which is likely to be slightly faster when accounting for the effect of all modes. The period of the high-frequency mode is equal to 24 fs, therefore the wavefunction is only in the vicinity of the crossing for a short time, compared to the low-frequency mode which has a period of 120 fs. As a result, only minor population transfer can occur during one period of the high-frequency mode, unlike the low-frequency case where the wavefunction develops significant CT character at these early timescales. This timescale-based argument is further supported by the fact that the energetics of the two excited states are almost identical along the high- and low-frequency modes: indeed, from the linear vibronic Hamiltonian of Eq. (7.7) it becomes obvious that along modes that share the same coupling matrix W_i and have similar λ values, the energy gap between the excited states is always similar and cannot explain the observed differences.

Moving our focus to the long-time dynamics and back to Fig. 7.9, for the excitation of the high-frequency mode, the dynamics are dissipative, with a gradual CT formation over approximately 2.5 ps. In the low-frequency case, the rapid CT formation at early times is followed by coherent oscillations, with a long damping timescale. Even at 4.5 ps coherent oscillations are still present. Despite the fact that the excess energy used to excite the high-frequency mode surpasses the one used to excite the

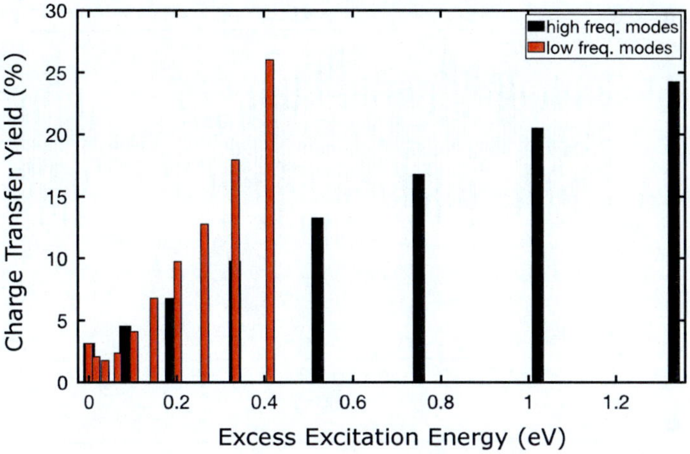

Fig. 7.11 Charge transfer yield at 2 ps following photoexcitation, for various values of the excitation energy, which is pumped into the system through low- or high-frequency modes. All the presented data refer to coherent excitation. For incoherent excitation, yields never surpass 5–6%

low-frequency mode by more than 1 eV, we see that the final yields of CT population are very close.

We would now like to further understand these long-time results. For both high- and low-frequency mode excitation, we plot in Fig. 7.11 the charge transfer yield at 2 ps of dynamics, for a range of values of the excess energy. The range of energies is limited by the maximum mode displacement Q we can simulate, related to the maximum bond dimension of the local Hilbert space of the MPS tensors (see methodology Sect. 7.2). This is why for the case of low-frequency modes, the maximum excess energy we can simulate is smaller than for high-frequency modes. However, we clearly see from Fig. 7.11 that exciting the low-frequency mode leads to significantly larger CT yields compared to the high-frequency case, at the same excess energy.

The qualitative differences between the long-time population dynamics of Fig. 7.9 (dissipative vs coherent oscillations), motivate us to plot the quantum coherence between the two exciton states in real-time. The coherences in Fig. 7.12b, c are clearly different, and we also plot the coherence for band-edge excitation for comparison, in Fig. 7.12a. For the excitation of a low-frequency mode, the coherence oscillations have an amplitude close to 0.4 (a maximally entangled state has a value of 0.5), and are only damped very slowly. When exciting a high-frequency mode, the coherence between the exciton states is more strongly damped and never goes beyond a value of 0.2, even at early times. The dissipation of coherence ceases at approximately 2.5 ps, when the charge transfer process of Fig. 7.9a is mostly complete. No such coherence dissipation is observed in Fig. 7.12a, where no excess energy is present. This leads us to the conclusion that the dissipation of coherence at long times is related to the vibrational relaxation which follows excess energy excitation. However, the differences between Fig. 7.12b, c point to the fact that this relaxation process can be

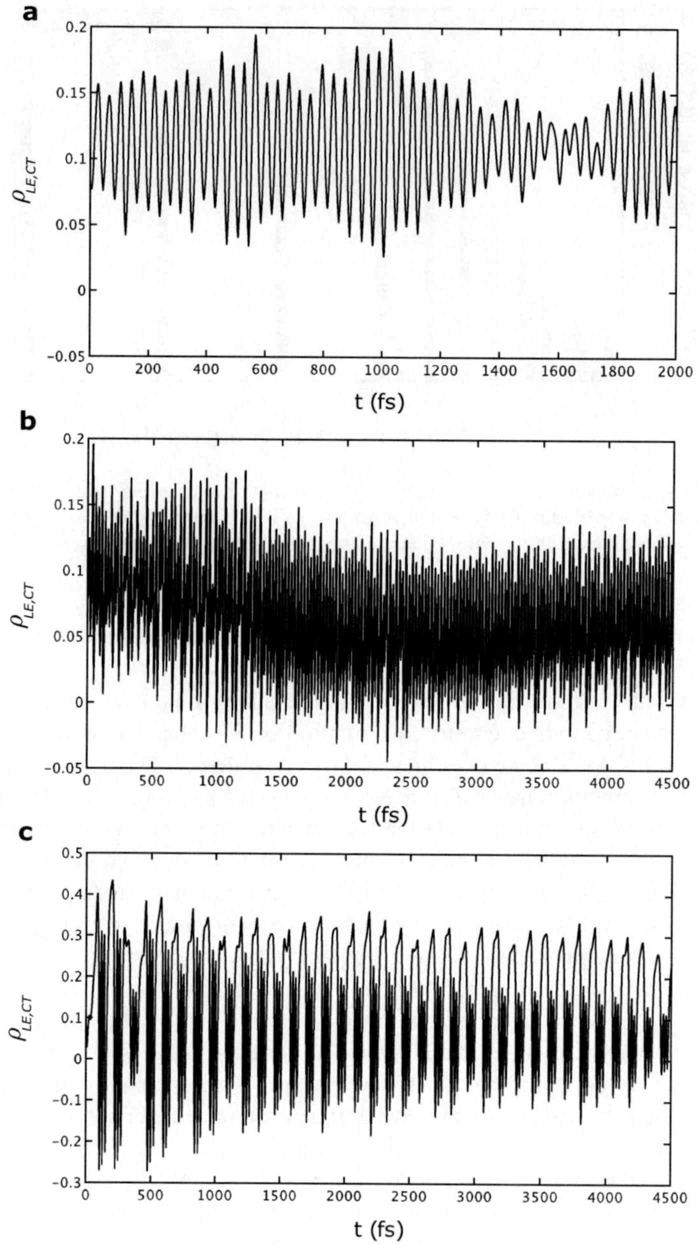

Fig. 7.12 Quantum coherence between the LE and CT states, for **a** band-edge excitation and excess energy excitation through, **b** a high- and **c** a low-frequency mode

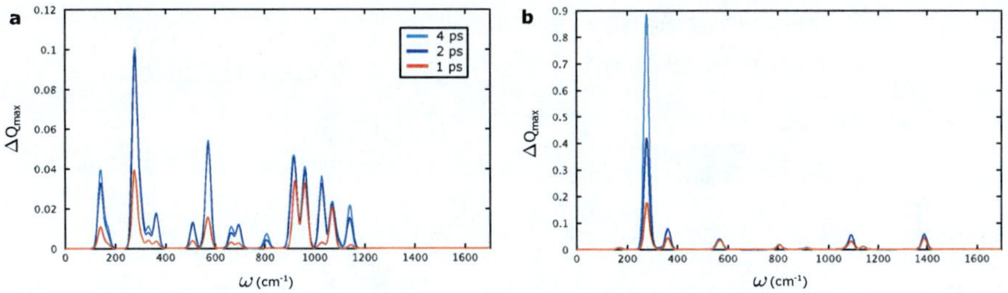

Fig. 7.13 Vibrational relaxation upon excitation of **a** the high-frequency and **b** the low-frequency mode. ΔQ_{max} denotes the change in mode amplitudes over the studied timescales

significantly different, depending on whether a low- or a high-frequency vibration was initially excited, as no significant coherence dissipation is present for the low-frequency case. The relaxation pathway of the system seems to have a direct impact on its ability to maintain a superposition state at long times, with the excitation of a low-frequency mode 'protecting' the quantum coherence between the two excitons, and allowing for a more efficient energy transfer. However the microscopic difference for this different coherence behaviour is still not clear at this stage.

A microscopic understanding of the different coherence-related properties may be obtained by considering the details of the vibrational relaxation process. Once we excite a high-frequency (i.e. high-energy) mode, the system will start to relax towards a Boltzmann distribution, where low-energy vibrations are predominantly displaced. Within our model picture, we expect the amplitude of low-frequency vibrations to increase with time. Indeed, as we find in Fig. 7.13a, the most significant increase in vibrational amplitude ΔQ_{max} over the course of 4 ps, occurs for low-frequency modes. As discussed when considering the early-time differences between exciting low-frequency and high-frequency modes, low-frequency vibrations facilitate more efficient crossings from the LE to the CT surface. Therefore, the gradual vibrational relaxation towards low-frequency modes coincides with the increase in the CT population observed in Fig. 7.9a. There is little change in the amplitudes of vibrations between 2 and 4 ps, suggesting that vibrational relaxation is mostly complete. Indeed the population of the CT state also reaches its maximum in these timescales.

Overall, there is a large number of vibrations towards which the excess energy relaxes once we excite a high-frequency mode. However, we note that our model only includes an implicit anharmonicity; the modes are only coupled to each other through the electronic system and not explicitly, making the transfer of the vibrational energy slower than if mode-mode energy transfer was included in our description. It is crucial in order to obtain realistic timescales for vibrational relaxation to include explicit anharmonicity in a model description. With the present approach, this would pose a significant challenge, as the coupling between vibrational modes would give rise to highly non-local interactions between MPS tensors. We therefore comment on the qualitative characteristics of the vibrational relaxation process, rather than the quantitative timescales.

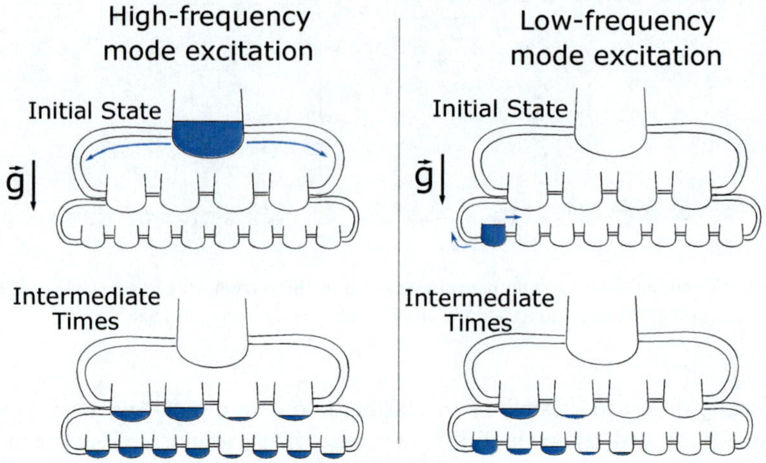

Fig. 7.14 A mechanical analogue to our model for vibrational relaxation. A system of jars of different sizes, filled with water in a gravitational field, and communicating through a system of pipes

The change in mode amplitudes looks very different, once excess energy excites a low-frequency vibration in Fig. 7.13b. The distribution of excited modes is now much more localised around low frequencies. The mode which gets strongly excited is degenerate with our low-frequency vibration of choice within $1\,\mathrm{cm}^{-1}$, so excess energy remains essentially 'trapped' (the modes of molecular dimers such as the studied one always go in pairs). Overall, only a small number of other oscillators are activated within 4 ps.

The above observations allow us to build a mechanical analogue, which rationalises the process of vibrational relaxation, and helps us understand the differences between low- and high-frequency mode excitation at long times. This is visualised in Fig. 7.14. Jars of various sizes are placed in a gravitational field, connected to each other through a system of pipes. Larger jars are placed higher up in the field, causing an efficient water flow towards lower-lying jars. By initially placing all of the available water in one of the high-lying jars, we end up at intermediate times (i.e. before timescales relevant at Boltzmann equilibrium), with a wide distribution of water over the low-lying jars.

On the other hand, if we place all of the available water in a low-lying jar, it can to some extent flow uphill through the pipe system, however this is rather inefficient because of gravity. Also, the flow to other low-lying jars is not at all affected by the presence of the gravitational field. Hence for intermediate timescales, we have a distribution of water which is much more localised compared to the case where we put all of the water in a high-lying jar. We emphasise here that both initial conditions will eventually lead to a Boltzmann equilibrium.

If we now substitute 'water' with energy, 'jars' of various sizes, with modes of different frequencies, 'pipes' with mode-mode coupling (anharmonicity), and 'gravity' with energy gradient, we get a model description of vibrational relaxation. Exciting a high-frequency mode results in relaxation towards lower energy vibrations, a process which is assisted by the existence of an energy gradient. Therefore, for intermediate timescales from adding excess energy to a high-frequency mode, we end up with a wide distribution of the energy over vibrational modes, predominantly of low-frequencies as seen in Fig. 7.13a. Once any of these oscillators is excited, it has a random phase, making it more and more difficult for the the total wavefunction to maintain any well-defined phase initially present. Therefore, the quantum coherence between the involved exciton states dissipates away. This becomes clear by comparing the slowly vanishing coherence of Fig. 7.12b to that of the band-edge excitation dynamics in Fig. 7.12a. These intermediate configurations resulting from high-frequency mode excitation are more incoherent, and approach the limit of incoherent excitation studied in Sect. 7.3.2.

For excitation of a low-frequency vibration, the limited pathways towards other modes lead to a more localised distribution of the excess energy at intermediate timescales, as visualised in Fig. 7.13b. It is therefore possible at these intermediate timescales to maintain the phase initially present in the system. Hence the quantum coherence between the exciton states is 'protected' for these intermediate timescales relevant to ultrafast processes, as seen in Fig. 7.12c.

7.4 Conclusions

In this chapter we have exhibited the implementation of a theoretical framework, which allows us to simulate the ultrafast dynamics of organic structures upon photoexcitation. Our tensor network approach allows us to perform a quantum mechanical time evolution of the vibrational wavefunction, which in turn makes it possible to visualise the real-time motion with atomistic level of detail. We have also developed methods to simulate the effects of excess energy excitation within the context of matrix product states.

We applied our methods to a covalently linked tetracene dimer, studying intramolecular charge transfer, a process which is endothermic in this system. Not only have we found that the final yield of charge transfer states strongly depends on the excitation energy of the light pulse, but also on the way this excess energy is introduced into the system. In particular, coherent excitation of a vibrational wavepacket opens a channel to charge transfer, whereas incoherent excitation only has a minor effect. We found the coherent excitation of low-frequency modes to be a much more efficient way of enhancing charge transfer, compared to the excitation of high-frequency modes. This is due to the different pathways of vibrational relaxation in the two cases; low-frequency vibrations have a more limited number of pathways towards a Boltzmann equilibrium, leading to a more localised distribution of excited oscillators in the intermediate timescales relevant for charge transfer. Thus the quantum coherence between the bright and charge transfer states is better preserved compared to the case of exciting high-frequency modes, where we found the dynamics to be more dissipative, approaching the limit of incoherent excitation.

Experimentally, the effect of mode-selective excitation on ultrafast processes has been demonstrated in the past [45, 46]. We believe that our study of the underlying mechanisms of vibrational relaxation following excess energy excitation could provide a guideline for future experimental work in this field, potentially focusing on exploiting specific vibrations for enhancing the efficiency of processes utilised in devices.

Our results for the studied two-level system are general, and will hold for any two-level system coupled with a bath of oscillators. As the system relaxes towards its Boltzmann equilibrium, the initial configuration, and in particular the part of the system where excess energy may initially may be localised, may have a strong effect on the properties at intermediate timescales. In particular, initial configurations with excess energy in low-frequency vibrations lie closer to the final Boltzmann distribution, reducing the number of pathways the system may follow towards equilibrium. Hence low-frequency modes preserve quantum coherence more efficiently, making them better candidates for the transfer of information.

Appendix 1: Convergence Tests

We study the convergence of the dynamics simulations in respect with the bond dimensions of the MPS. There are two separate bond dimensions to consider: The first is the intra-chain bond dimension (D), which encodes the entanglement between the environment tensors, while the second (D_h) refers to the system-environment coupling (see Fig. 7.2). We use the yield of the CT state at 2 ps as an indicator for convergence. We study the convergence of dynamics following band-edge excitation of the CT state (akin to Fig. 7.6b), in Fig. 7.15a. The values which were used in the main part of the chapter are highlighted. We have also examined the convergence in the case of adding excess energy to the high-frequency vibrations, displacing it by $Q = 2$, and the results are shown in Fig. 7.15b.

We have also tested the convergence of the dynamics following excess energy excitation of the high-frequency mode, in respect with the time step of the simulation. The results are shown in Fig. 7.16. We use a time-step of 0.67 fs, despite the fact that this value is not perfectly converged. This allows us to significantly speed-up our calculations, and investigate the results at longer timescales. Compared to the smallest value for the time step of the simulations, the error in the CT yield is smaller than 0.3%.

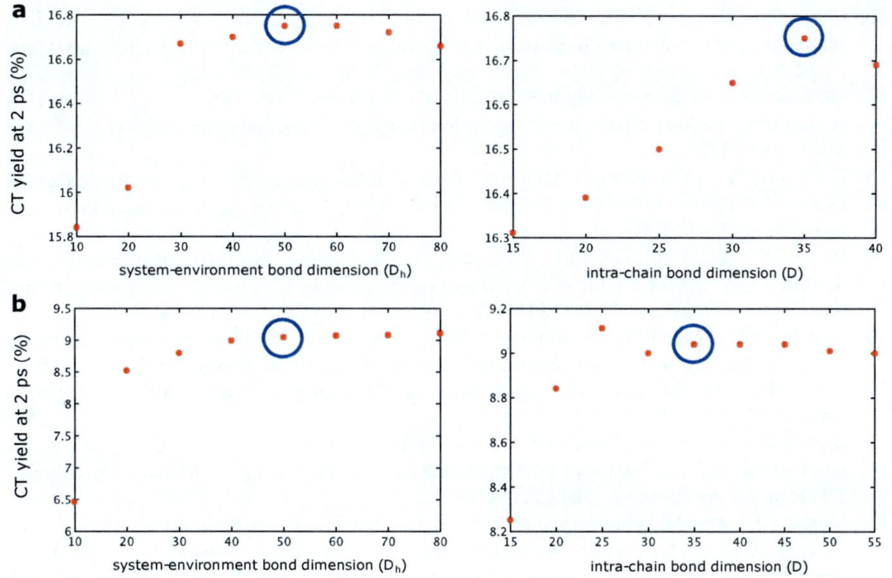

Fig. 7.15 Convergence of CT populations at 2 ps with respect to bond dimensions, for **a** band-edge and **b** excess-energy excitation

Fig. 7.16 Convergence of CT populations at 2 ps with respect to simulation time step, for the case of excess energy excitation of the high-frequency mode

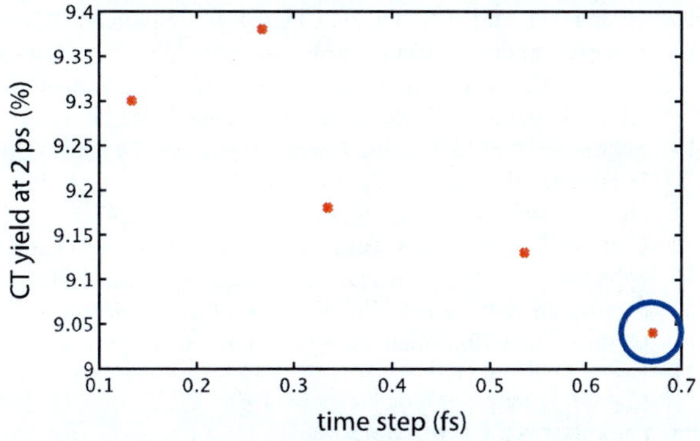

References

1. Alvertis AM, Schröder FAYN, Chin AW (2019) Non-equilibrium relaxation of hot states in organic semiconductors: impact of mode-selective excitation on charge transfer. J Chem Phys 151:084104
2. Blankenship R (2014) Molecular mechanisms of photosynthesis. Wiley-Blackwell
3. van Amerongen H, Valkunas L, van Grondelle R (2000) Photosynthetic excitons World Scientific
4. Kippelen B, Bredas Jean-Luc (2008) Organic photovoltaics. Energy Environ Sci 2:251
5. O'Reilly EJ, Olaya-Castro A (2014) Non-classicality of the molecular vibrations assisting exciton energy transfer at room temperature. Nat Commun 5:3012

6. Cabanillas-Gonzalez J, Grancini G, Lanzani G (2011) Pump-probe spectroscopy in organic semiconductors: monitoring fundamental processes of relevance in optoelectronics. Adv Mater 23:5468
7. Gélinas S, Rao A, Kumar A, Smith SL, Chin AW, Clark J, van der Poll TS, Bazan GC, Friend RH (2014) Ultrafast long-range charge separation in organic semiconductor photovoltaic diodes. Science 343:512
8. Falke SM, Rozzi CA, Brida D, Maluri M, Amato M, Sommer E, De Sio A, Rubio A, Cerullo G, Molinari E, Lienau C (2014) Coherent ultrafast charge transfer in an organic photovoltaic blend. Science 344:1001
9. Wong CY, Penwell SB, Cotts BL, Noriega R, Hao W, Ginsberg NS (2013) Revealing exciton dynamics in a small-molecule organic semiconducting film with subdomain transient absorption microscopy. J Phys Chem C 117:22111
10. Wan Y, Guo Z, Zhu T, Yan S, Johnson J, Huang Libai (2015) Cooperative singlet and triplet exciton transport in tetracene crystals visualized by ultrafast microscopy. Nat Chem 7:785
11. Smith SL, Chin AW (2015) Phonon-assisted ultrafast charge separation in the PCBM band structure. Phys Rev B 91:201302
12. Bakulin AA, Morgan SE, Kehoe TB, Wilson MWB, Chin AW, Zigmantas D, Egorova D, Rao A (2016) Real-time observation of multiexcitonic states in ultrafast singlet fission using coherent 2D electronic spectroscopy. Nat Chem 8:16
13. Musser AJ, Liebel M, Schnedermann C, Wende T, Kehoe TB, Rao A, Kukura P (2015) Evidence for conical intersection dynamics mediating ultrafast singlet exciton fission. Nat Phys 11:352
14. Tempelaar R, Reichman DR (2017) Vibronic exciton theory of singlet fission. I. Linear absorption and the anatomy of the correlated triplet pair state. J Chem Phys 146:174703
15. Bakulin AA, Rao A, Pavelyev VG, van Loosdrecht PHM, Pshenichnikov MS, Niedzialek D, Cornil J, Beljonne D, Friend RH (2012) The role of driving energy and delocalized states for charge separation in organic semiconductors. Science 335:1340
16. Grancini G, Maiuri M, Fazzi D, Petrozza A, Egelhaaf HJ, Brida D, Cerullo G, Lanzani G (2013) Hot exciton dissociation in polymer solar cells. Nat Mater 12:29
17. Brédas J-L, Sargent EH, Scholes GD (2017) Photovoltaic concepts inspired by coherence effects in photosynthetic systems. Nat Mater 16:35
18. Scholes GD, Fleming GR, Chen LX, Aspuru-Guzik A, Buchleitner A, Coker DF, Engel GS, Grondelle RV, Ishizaki A, Jonas DM et al (2017) Using coherence to enhance function in chemical and biophysical systems. Nature 543:647
19. Creatore C, Parker MA, Emmott S, Chin AW (2013) Efficient biologically inspired photocell enhanced by delocalized quantum states. Phys Rev Lett 111:253601
20. Sarovar M, Ishizaki A, Fleming GR, Birgitta Whaley K (2010) Quantum entanglement in photosynthetic light-harvesting complexes. Nat Phys 6:462
21. Bacon D, Brown KR, Birgitta Whaley K (2001) Coherence-preserving quantum bits. Phys Rev Lett 87:247902
22. Chin AW, Prior J, Rosenbach R, Caycedo-Soler F, Huelga SF, Plenio MB (2013) The role of non-equilibrium vibrational structures in electronic coherence and recoherence in pigment-protein complexes. Nat Phys 9:113
23. Roden J, Strunz WT, Birgitta Whaley K, Eisfeld A (2012) Accounting for intra-molecular vibrational modes in open quantum system description of molecular systems. J Chem Phys 137:204110
24. Bittner ER, Silva C (2014) Noise-induced quantum coherence drives photo-carrier generation dynamics at polymeric semiconductor heterojunctions. Nat Commun 5:3119
25. Berkelbach TC, Hybertsen MS, Reichman DR (2013) Microscopic theory of singlet exciton fission. II. Application to pentacene dimers and the role of superexchange. J Chem Phys 138:114103
26. Mirjani F, Renaud N, Gorczak N, Grozema FC (2014) Theoretical investigation of singlet fission in molecular dimers: the role of charge transfer states and quantum interference. J Phys Chem C 118:14192

27. Zarea M, Powell D, Renaud N, Wasielewski MR, Ratner MA (2013) Decoherence and quantum interference in a four-site model system: mechanisms and turnovers. J Phys Chem B 117:1010
28. Zheng Yu J, Xie SJ, Lan Z (2016) Ultrafast nonadiabatic dynamics of singlet fission: quantum dynamics with the multilayer multiconfigurational time-dependent Hartree (ML-MCTDH) method. J Phys Chem C 120:1375
29. Binder R, Lauvergnat D, Burghardt I (2018) Conformational dynamics guides coherent exciton migration in conjugated polymer materials: first-principles quantum dynamical study. Phys Rev Lett 120:227401
30. Monahan NR, Sun D, Tamura H, Williams KW, Xu B, Zhong Y, Kumar B, Nuckolls C, Harutyunyan AR, Chen G, Dai HL, Beljonne D, Rao Y, Zhu XY (2017) Dynamics of the triplet-pair state reveals the likely coexistence of coherent and incoherent singlet fission in crystalline hexacene. Nat Chem 9:341
31. Monserrat B, Engel EA, Needs RJ (2015) Giant electron-phonon interactions in molecular crystals and the importance of nonquadratic coupling. Phys Rev B 92:140302(R)
32. Shi YY, Duan LM, Vidal G (2006) Classical simulation of quantum many-body systems with a tree tensor network. Phys Rev A 74:022320
33. Schollwöck U (2011) The density-matrix renormalization group in the age of matrix product states. Ann Phys 326:96
34. Schröder FAYN, Turban DHP, Musser AJ, Hine NDM, Chin AW (2019) Tensor network simulation of multi-environmental open quantum dynamics via machine learning and entanglement renormalisation. Nat Commun 10:1062
35. Pino JD, Schröder FAYN, Chin AW, Feist J, Garcia-Vidal FJ (2018) Tensor network simulation of polaron-polaritons in organic microcavities. Phys Rev B 98:165416
36. Pino JD, Schröder FAYN, Chin AW, Feist J, Garcia-Vidal FJ (2018) Tensor network simulation of non-Markovian dynamics in organic polaritons. Phys Rev Lett 121:227401
37. Wang H, Thoss M (2003) Multilayer formulation of the multiconfiguration time-dependent Hartree theory. J Chem Phys 119:1289
38. Vandewal K, Albrecht S, Hoke ET, Graham KR, Widmer J, Douglas JD, Schubert M, Mateker WR, Bloking JT, Burkhard GF, Sellinger A, Fréchet JMJ, Amassian A, Riede MK, McGehee MD, Neher D, Salleo A (2014) Efficient charge generation by relaxed charge-transfer states at organic interfaces. Nat Mater 13:63
39. Kurpiers J, Ferron T, Roland S, Jakoby M, Thiede T, Jaiser F, Albrecht S, Janietz S, Collins BA, Howard IA, Neher D (2018) Probing the pathways of free charge generation in organic bulk heterojunction solar cells. Nat Commun 9:2038
40. Musser AJ, Al-Hashimi M, Maiuri M, Brida D, Heeney M, Cerullo G, Friend RH, Clark J (2013) Activated singlet exciton fission in a semiconducting polymer. J Am Chem Soc 135:12747
41. Paleček D, Edlund P, Westenhoff S, Zigmantas D (2017) Quantum coherence as a witness of vibronically hot energy transfer in bacterial reaction center. Sci Adv 3:e1603141
42. Beratan DN, Waldeck DH (2016) DNA charge transfer: hot holes break the speed limit. Nat Chem 8:992
43. Tamura H, Martinazzo R, Ruckenbauer M, Burghardt I (2012) Quantum dynamics of ultrafast charge transfer at an oligothiophene—fullerene heterojunction. J Chem Phys 137:22A540
44. Kloss B, Reichman DR, Tempelaar R (2019) Multiset matrix product state calculations reveal mobile Franck-Condon excitations under strong Holstein-type coupling. Phys Rev Lett 123:126601
45. Bakulin AA, Lovrincic R, Yu X, Selig O, Bakker HJ, Rezus YLA, Nayak PK, Fonari A, Coropceanu V, Brédas JL, Cahen D (2015) Mode-selective vibrational modulation of charge transport in organic electronic devices. Nat Commun 6:7880
46. Delor M, Keane T, Scattergood PA, Sazanovich IV, Greetham GM, Towrie M, Meijer AJHM, Weinstein JA (2015) On the mechanism of vibrational control of light-induced charge transfer in donor-bridge-acceptor assemblies. Nat Chem 7:689
47. Guo C, Weichselbaum A, von Delft J, Vojta M (2012) Critical and strong-coupling phases in one- and two-bath spin-boson models. Phys Rev Lett 108:160401

48. Schröder FAYN, Chin AW (2016) Simulating open quantum dynamics with time-dependent variational matrix product states: towards microscopic correlation of environment dynamics and reduced system evolution. Phys Rev B 93:075105
49. Chin AW, Rivas Á, Huelga SF, Plenio MB (2010) Exact mapping between system-reservoir quantum models and semi-infinite discrete chains using orthogonal polynomials. J Math Phys 51:092109
50. Prior J, Vega I, Chin AW, Huelga SF, Plenio MB (2013) Quantum dynamics in photonic crystals. Phys Rev A 87:013428
51. Prior J, Chin AW, Huelga SF, Plenio MB (2010) Efficient simulation of strong system-environment interactions. Phys Rev Lett 105:050404
52. Arthur D, Vassilvitskii S (2007) K-Means++: the advantages of careful seeding. In: Proceedings of the eighteenth annual ACM-SIAM symposium on Discrete algorithms, p 1027
53. Van Der Maaten L, Hinton G, van der Maaten GH (2008) Visualizing data using t-SNE. J Mach Learn Res 9:2579
54. Haegeman J, Lubich C, Oseledets I, Vandereycken B, Verstraete F (2016) Unifying time evolution and optimization with matrix product states. Phys Rev B 94:165116
55. Vidal G (2007) Entanglement renormalization. Phys Rev Lett 99:220405
56. Barnett SM, Radmore PM (2005) Methods in theoretical quantum optics. Oxford Series in optical and imaging science, vol 15. Clarendon, Oxford
57. Smith MB, Michl J (2010) Singlet fission. Chem Rev 110:6891
58. Rao A, Friend RH (2017) Harnessing singlet exciton fission to break the Shockley–Queisser limit. Nat Rev Mater 2:17063
59. Wilson MWB, Rao A, Johnson K, Gélinas S, Pietro RD, Clark J, Friend RH (2013) Temperature-independent singlet exciton fission in tetracene. J Am Chem Soc 135:16680
60. Stern HL, Cheminal A, Yost SR, Broch K, Bayliss SL, Chen K, Tabachnyk M, Thorley K, Greenham N, Hodgkiss JM, Anthony J, Head-Gordon M, Musser AJ, Rao A, Friend RH (2017) Vibronically coherent ultrafast triplet-pair formation and subsequent thermally activated dissociation control efficient endothermic singlet fission. Nat Chem 9:1205
61. Lukman S, Chen K, Hodgkiss JM, Turban DHP, Hine NDM, Dong S, Wu J, Greenham NC, Musser AJ (2016) Tuning the role of charge-transfer states in intramolecular singlet exciton fission through side-group engineering. Nat Commun 7:13622
62. Pople JA (1953) Electron interaction in unsaturated hydrocarbons. Trans Faraday Soc 49:1375
63. Pariser R, Parr RG (1953) A semi-empirical theory of the electronic spectra and electronic structure of complex unsaturated molecules. I. J Chem Phys 21:466
64. Tavan P, Schulten K (1987) Electronic excitations in finite and infinite polyenes. Phys Rev B 36:4337
65. Aryanpour K, Shukla A, Mazumdar S (2015) Theory of singlet fission in polyenes, acene crystals, and covalently linked acene dimers. J Phys Chem C 119:6966
66. Njegic B, Gordon MS (2006) Exploring the effect of anharmonicity of molecular vibrations on thermodynamic properties. J Chem Phys 125:224102

Chapter 8
Molecular Movie of Ultrafast Singlet Exciton Fission

The results presented in this chapter have been published in Ref. [1] (https://doi.org/10.1038/s41467-019-12220-7): Christoph Schnedermann, Antonios M. Alvertis, Torsten Wende, Steven Lukman, Jiaqi Feng, Florian A.Y.N. Schröder, David H.P. Turban, Jishan Wu, Nicholas D.M. Hine, Neil C. Greenham, Alex W. Chin, Akshay Rao, Philipp Kukura, Andrew J. Musser. A Molecular Movie of Ultrafast Singlet Fission. Nature Communications, 10:4207, 2019, under a Creative Commons license: https://creativecommons.org/licenses/by/4.0/ . Changes to the text and figures were made compared to the published article.
The experimental work presented in this chapter was not undertaken by the author of this thesis, but by a number of collaborators who are listed as co-authors in the aforementioned publication: Cristoph Schnedermann, Torsten Wende, Steven Lukman, Jiaqi Feng, Jishan Wu, Neil C. Greenham, Akshay Rao, Philipp Kukura and Andrew J. Musser. The measurements and their interpretation are an integral part of this combined theoretical-experimental study, and are hence presented here for completeness, along with the associated experimental methodology.

In this chapter we employ the tree tensor network state (TTNS) formalism presented in Chap. 7 in order to study the full quantum dynamics of the process of singlet exciton fission in a covalent pentacene dimer with 108 atoms. We monitor the real-time displacement of the molecular normal modes, rigorously comparing to experiment and confirming the accuracy of our non-perturbative approach. This allows us to reveal the full microscopic mechanism of singlet fission in the studied system, and we find that two distinct kinds of normal modes coordinate in a precise manner in order to allow this process to occur efficiently. We therefore highlight the critical effect of vibrations on real-time exciton dynamics, as well as the potential of the TTNS framework for successfully describing exciton evolution in large, experimentally-accessible systems.

© The Author(s), under exclusive license to Springer Nature Switzerland AG 2021

A. M. Alvertis, *On Exciton–Vibration and Exciton–Photon Interactions in Organic Semiconductors*, Springer Theses,
https://doi.org/10.1007/978-3-030-85454-6_8

8.1 Introduction

The coupling between vibrational and electronic degrees of freedom after photon absorption defines photochemical reaction pathways and guides processes such as charge and exciton generation [2, 3], transport [4, 5] and recombination [6] as well as photoisomerisation [7] and bond-dissociation [8]. The initial photoreactivity after photoexcitation is governed by ultrafast processes, including a correlated evolution along vibrational coordinates and their associated electronic states on the reaction coordinate. As a result, these processes cannot be described in the framework of the Born–Oppenheimer approximation [2, 9–11]. Despite remarkable progress in the optical manipulation of vibrational and electronic states [4, 6] and the identification of vibronically coherent processes [2, 9, 12, 13], the precise molecular mechanisms and associated structural changes remain largely elusive and subject to competing interpretations.

This uncertainty stems from a disparity between experimental and theoretical methods. Structurally-sensitive experimental techniques that can access the earliest photoreactive transformations are often only available for large and complex molecular systems, while accurate first-principles computational modelling for such non-Born–Oppenheimer dynamic processes is only affordable for much smaller models [14, 15]. Consequently, even if experimental structural information is available, it can rarely be accurately projected onto the molecular origin of the crucial coupling and tuning modes involved, preventing identification of the operative reaction mechanisms [16, 17]. Developing a detailed molecular understanding of such complex photoinduced processes is crucial, however, to provide rational design criteria for improved functional materials, for instance for organic optoelectronics and molecular photocatalysts.

To this end, novel theoretical methods that can address complex molecular systems are critical. The experimental validation of such theories must go beyond simple population dynamics—too coarse a figure of merit—and explicitly include structural configuration changes after photoexcitation [18]. In this chapter, we demonstrate the power of such a combined approach as applied to the model process of singlet fission, i.e. the conversion of a photoexcited singlet exciton (S_1) into two triplet excitons via a correlated triplet pair (^1TT) intermediate [19].

Singlet fission is a classic example of an ultrafast process in which the molecular mechanism can only be inferred due to the lack of experimentally and theoretically comparable data sets, despite extensive study over the past decade [20–24]. In recent years, it has been demonstrated in several thin-film singlet fission systems that the initial $S_1 - {}^1TT$ conversion is vibrationally coherent [12, 13, 25–27]. Other studies using structurally sensitive techniques have also found that the key electronic processes in singlet fission are linked to inter- and intra-molecular motions [28, 29]. However, the structural complexity of these systems precluded direct interpretation in terms of specific motions and their role in the reaction. Within the theoretical community, studies have shown that the typical vibronic couplings in singlet fission materials are strong (10's to 100's meV) and thus require non-perturbative methods to be accurately described [26, 30, 31], and the same is true in many other molecular systems [32]. As a result, advanced simulation techniques have been applied to elucidate the varied roles of ultrafast and non-equilibrium environmental dynamics on fission [33–36]. There is growing recognition that singlet fission, like the majority of ultrafast (< 10 ps) processes, is intimately coupled to nuclear dynamics [23, 30, 31, 37–43]. Nonetheless there is no clear determination of what motions drive the process, how this coupling occurs, or indeed whether the reported vibrational coherence is important in achieving a high reaction yield or simply a consequence of the ultrafast nature of the reaction with no functional importance. These problems are typical of the more general study of non-Born–Oppenheimer dynamics, and they constitute a key bottleneck in materials understanding and design.

One example system in which such ultrafast structural changes are expected to play a significant role in the electronic dynamics is the large (108-atom) and complex pentacene dimer DP-Mes (Fig. 8.1a). Like many dimers of pentacene [44–46], DP-Mes is capable of sub-ps intramolecular singlet fission [47, 48]. The reaction rates and ^{1}TT yields depend strongly on solvent environment [47, 48], indicating that fission is mediated by coupling to higher-energy charge-transfer states [24, 39, 40, 43] (Fig. 8.1b). In the previous chapter, we presented a full quantum-mechanical algorithm based on Tree Tensor Network states (TTNS) to simulate the full structural dynamics of charge transfer, which has also been applied in the context of singlet fission in DP-Mes [49]. This TTNS method is largely based on the formalism of matrix product states and tree tensor states, which have been successfully applied to condensed matter problems [50–55] and—more recently—to the dynamics of open quantum systems [56–62]. In a previous work on DP-Mes [49], a combination of machine learning and entanglement renormalisation techniques was employed to capture the non-perturbative and non-Markovian physics arising from strong coupling to a large number of vibrational modes. A recent theoretical study highlighted the breakdown of perturbative approaches in predicting singlet fission rates in pentacene dimer systems [63], further emphasising the fact that non-perturbative approaches such as the TTNS method [49] are necessary to describe fission dynamics. The DP-Mes simulations in [49] confirmed that singlet fission is mediated by superexchange and is driven by a chain of cooperative vibronic processes involving modes of different timescales and symmetries.

In this chapter, ultrafast excited-state vibrational spectroscopy is employed to probe the transfer of vibrational wavepackets from the initially photoexcited S_1 state to the ^{1}TT state in DP-Mes. We build on previous TTNS simulations presented in [49] in order to map the structural dynamics to real experimental observables. This exact quantum treatment is uniquely suited to describe systems such as DP-Mes, which have strong vibronic couplings (up to 0.3 eV). The model considers 5 excited electronic states—two symmetry-adapted singlet states, two charge-transfer states and ^{1}TT—and 252 strongly coupled vibrational modes whose properties are obtained from *ab initio* electronic structure techniques [47]. The simulations closely reproduce both the frequencies and the intensities of the experimentally retrieved vibrational coherence signatures. The remarkable structural agreement between theory and experiment enables us to reconstruct the real-space structural motion associated

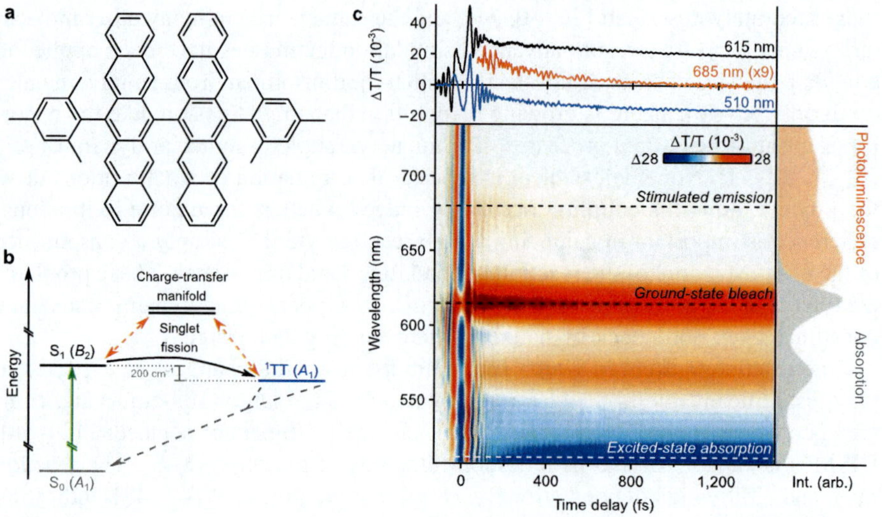

Fig. 8.1 **a** Chemical structure of DP-Mes. The molecule assumes an orthogonal configuration in its ground state. **b** DP-Mes electronic states and photophysics. Singlet fission from S_1 to ^1TT is mediated by coupling (orange) to a manifold of charge-transfer states which are not directly populated, facilitating the overall marginally exothermic ($\approx 200\,\text{cm}^{-1}$) process [47, 48]. Symmetries of the relevant excited states are indicated in parentheses [49]. **c** Differential transmission map of a DP-Mes thin film following excitation with a 13 fs pulse centred at 550 nm, at room temperature. The absorption (grey) and photoluminescence (orange) spectra are shown to guide spectral assignment. The primary spectral features of S_1 (stimulated emission), ^1TT (excited-state absorption) and total excited-state population (ground-state bleach) are indicated by dashed lines. The associated transient kinetics are shown above, with the weak stimulated emission trace scaled by a factor of 9 for clarity. The superimposed oscillatory modulations correspond to vibrationally coherent wavepackets formed by impulsive excitation

with the vibronic dynamics as a real-time movie, giving a fully quantum visualisation of the process and enabling determination of the nature of the critical coupling and tuning modes driving this coherent ultrafast reaction. Our results provide a refined atomistic picture of the molecular mechanism of singlet fission and allow us to visualise ultrafast quantum dynamics in an experimentally verifiable way.

8.2 Methodology

In order to model the real-time dynamics of singlet fission in DP-Mes we employ the tensor network formulation outlined in Chap. 7. As in Chap. 7, we employ the linear vibronic Hamiltonian:

Table 8.1 Electronic states of DP-Mes relevant to singlet fission, alongside their computed energy and symmetry group

State	Energy (eV)	Irreducible representation
TT	1.83	A_1
LE_B (S_1)	2.07	B_2
LE_A	2.20	A_1
CT_B	2.75	A_2
CT_A	2.76	B_1

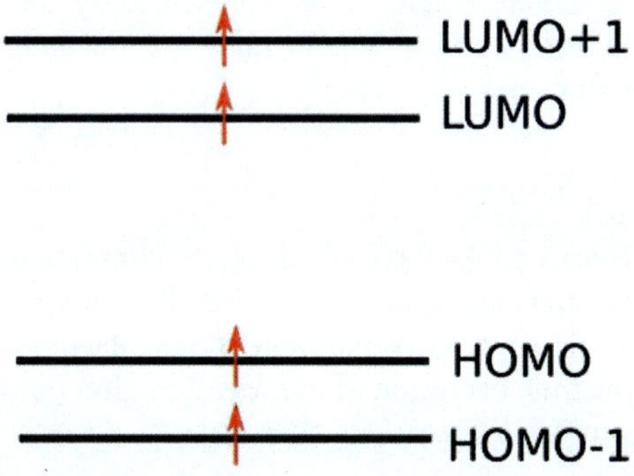

Fig. 8.2 The ground state quintet (i.e. $S = 2$ state) used to approximate TT

$$H_{star} = H_{el} + \sum_i \overline{W_i} \sum_k \lambda_{i,k} \frac{b_{i,k}^\dagger + b_{i,k}}{\sqrt{2}} + H_{c,i}, \tag{8.1}$$

where once again we generate clusters (i) of vibrational modes (k). In particular, we employ the parametrisation of a previous study on DP-Mes [49]. Since the structure of DP-Mes exhibits a D_{2d} symmetry, the relevant electronic states for singlet fission transform as the irreducible representation of the point group, which we summarise in Table 8.1. Similar to Chap. 7, we denote local excitons by LE and charge transfer states by CT. The energies of these excitons are obtained in [49] using time-dependent density functional theory. Regarding the final product of singlet fission, i.e. the singlet TT state ^{1}(TT), its energy is approximated as that of the spin-two ground state, i.e. the ground state quintet ^{5}(TT) which is visualised in Fig. 8.2. The energy difference between the singlet and quintet configurations of the TT state is usually not larger than 10 meV [64, 65], we therefore expect this to provide a reasonable approximation.

The vibrational modes of DP-Mes can also be classified according to their symmetry, and the generated clusters loosely correspond to the various irreducible representations of the point group. The form of the calculated coupling matrices $\overline{W_i}$ that appear in Eq. (8.1) (see Appendix 2), furthermore allows us to characterise the diagonal modes of A_1 symmetry as tuning modes, since these only serve to tune the energies of the electronic states. At the same time, off-diagonal modes of A_2, B_1 and B_2 symmetry act as coupling modes of the singlet fission reaction, since they couple the various electronic states to each other. In the model Hamiltonian of Eq. (8.1) we include 252 out of the 318 vibrations of DP-Mes. We exclude the very slow modes with frequencies below 110 cm^{-1} as these are highly anharmonic. Likewise, we exclude the very high-frequency C–H stretches above 1680 cm^{-1} as they are expected to make little contribution to the overall dynamics and they are beyond the scope of the experimental technique employed in this chapter. The molecular vibrations are calculated at the cc-pVDZ, B3LYP level of DFT, using the NWChem software [66].

In Chap. 7 we outlined the orthogonal polynomial mapping U applied to the physical vibrational normal modes of a system in order to bring it to the linear architecture that can be represented by a matrix product state (Fig. 7.2). This results in a set of new 'chain modes' b_k, which are linear combinations of the original modes a_k:

$$b_i = \sum_k U_{ik} a_k. \tag{8.2}$$

Here we take the critical step to compute the real-time displacement of the *physical* normal modes, in order to directly compare with experimental Raman measurements. To do so, we first compute the displacement Δ of all chain modes n at each step of the time evolution of the wavefunction (which is expanded into a network of tensors similar to Chap. 7):

$$\langle \Delta \rangle_{n,\text{chain}} = \frac{\langle b_n^\dagger + b_n \rangle}{\sqrt{2}}. \tag{8.3}$$

The orthogonal polynomial transformation U of the molecular vibrations into the chain modes may be inversed to produce the operators corresponding to the physical normal modes of the system:

$$a_k = \sum_i U_{ki}^* b_i. \tag{8.4}$$

Since U is a unitary matrix for each chain, we deduce:

$$\langle \Delta \rangle_{n,\text{mode}} = \sum_l U_{kl}^{-1} \langle \Delta \rangle_{n,\text{chain}} \tag{8.5}$$

The recorded impulsive Raman intensity depends on the dimensionless displacement parameter, Δ, of the state-specific vibrational normal modes of DP-Mes according to Eq. (8.6), where ω denotes vibrational frequency:

$$I \propto \Delta^2 \omega^2. \tag{8.6}$$

We therefore expanded the TTNS approach by projecting the calculated time-dependent displacements of the tensor states onto the molecular normal-mode vibrations of DP-Mes determined by density-functional theory. In reality, there are a few more subtle points when it comes to drawing a direct comparison between the theoretical and experimental data, which are associated with the precise way that the experimental measurements are performed. Here we have presented the basic principles, and more detail is provided in Appendix 4. Details on the experimental methods employed to study singlet fission in DP-Mes are given in Appendix 1.

8.3 Results

8.3.1 *Vibrationally Coherent Singlet Fission*

Pump-probe spectroscopy with a time resolution of $\approx$ 13 fs has been used to track the singlet fission process in DP-Mes (Fig. 8.1c). The strong positive bands ($\Delta T/T > 0$) match the ground-state absorption peaks (grey spectrum) and can be attributed to a ground-state bleach signal. At early time delays (< 0.7 ps) this bleach is spectrally overlaid with the characteristic stimulated emission ($\Delta T/T > 0$) of S_1 in the range 625–725 nm, in agreement with the very short-lived [67] photoluminescence (orange spectrum). The S_1 state decays concurrently with the rise of a distinctive excited-state absorption ($\Delta T/T < 0$) peaked at $\approx$ 515 nm previously assigned to ^{1}TT [47, 67], with a time constant of 320 fs. These dynamics are consistent with previous reports using low-power narrow-band excitation and reveal highly efficient ($> 90\%$) singlet fission [47, 67].

The electronic population dynamics show distinct oscillatory modulations throughout the probed spectral window, which report on impulsively generated vibrational wavepacket motion on both ground- and excited-state potential energy surfaces [68–71]. This vibrational coherence can be isolated through subtraction of the slower electronic dynamics at each detection wavelength followed by Fourier transformation, which yields the impulsive Raman spectrum detected at each probe wavelength (Fig. 8.3a).

Three distinct spectral regions of vibrational coherence activity are discernible in Fig. 8.3a, matching the ground-state bleach, stimulated emission and excited-state absorption features identified above (Fig. 8.1c). The impulsive Raman spectrum in the ground-state bleach region (Fig. 8.3b, black) exhibits several peaks in the high-frequency region at 1153, 1210 and 1372 cm^{-1} as well as an intense low-frequency mode at 263 cm^{-1} and a weaker mode at 785 cm^{-1}. The Raman spectrum is in good agreement with a ground-state impulsive Raman spectrum measured separately as a reference (Fig. 8.3b, grey spectrum, experimental details in Appendix 1), indicating that this spectral region is dominated by vibrational activity on the ground state, S_0. We attribute the difference in relative peak intensities between the two S_0 spectra, especially in the low-frequency region, to the different pump resonance enhancement conditions employed [72].

The impulsive Raman spectrum obtained in the stimulated emission region (Fig. 8.3b, red) reveals similar peak positions to the ground-state spectrum at 263, 608, 785, 1160, 1198 and 1372 cm^{-1}. These are slightly shifted (< 10 cm^{-1}) and exhibit markedly different intensity profile, especially in the high-frequency region. Based on these subtle differences, this spectrum is tentatively assigned to the excited

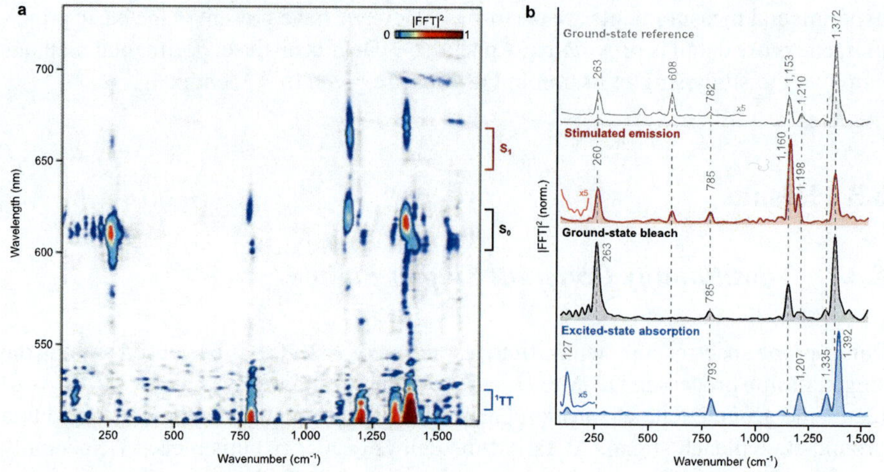

Fig. 8.3 **a** Fast Fourier transform (FFT) power map of DP-Mes thin film following impulsive photoexcitation into S_1. **b** Impulsive Raman spectra for different probe spectral components, integrated as shown by brackets in panel **a**: stimulated emission—645–665 nm (red, S_1), ground-state bleach—605–625 nm (black, S_0), excited-state absorption—515–525 nm (blue, ^{1}TT). The off-resonant impulsive Raman spectrum of DP-Mes in S_0 averaged over probe wavelengths from 605 to 625 nm is shown for comparison (grey, details in Appendix 1). Dashed vertical lines highlight the difference between ground-state vibrational modes and those identified in ^{1}TT. All spectra are normalised to the strongest peak

S_1 state, albeit with underlying ground-state contributions preventing unambiguous assignment [73].

In the excited-state absorption band (Fig. 8.3b, blue) the impulsive Raman spectrum reveals pronounced differences from the ground-state region both in relative peak intensities and frequencies, allowing confident assignment to the ^{1}TT state. Compared to the S_0 and S_1 Raman spectra (Fig. 8.3b, black and red), the ^{1}TT Raman spectrum exhibits higher-frequency peak positions at 793, 1335 and 1392 cm^{-1} as well as novel bands at 127 and 1126 cm^{-1}. The observation of vibrational coherence in ^{1}TT, which is not directly photoexcited, has important mechanistic consequences. It demonstrates that the molecular vibrations initiated on photoexcitation are precisely synchronised with the change in electronic state and accompanying shifts in frequency, ruling out stochastic hopping or tunneling between S_1 and ^{1}TT surfaces. Such a phenomenon could be explained through smooth evolution along a simple adiabatic potential energy surface from S_1-rich to ^{1}TT-rich character [22], but this is not the case in DP-Mes: S_1 and ^{1}TT exhibit negligible mixing at the orthogonal Franck-Condon point [47, 67] and recent TTNS simulations indicate the fission process is driven by non-adiabatic coupling through vibronic superexchange [49]. Instead, the finding of vibrational coherence across the full vibrational fingerprint region in ^{1}TT requires a vibrationally coherent process in which the photogenerated vibrational wavepackets on S_1 are transferred to ^{1}TT [9]. Earlier studies of intermolecular singlet fission in TIPS-pentacene and other acene films reported a similar behaviour [12, 26, 74]. The intramolecular process in DP-Mes thus follows the same mechanism identified for intermolecular fission in TIPS-pentacene, even though it proceeds substantially slower in the dimer system (320 fs vs. 80 fs).

8.3.2 *Identification of Transferred Coherence*

To elucidate the origin of the different contributions in the ^{1}TT vibrational coherence spectrum and how it is affected by the $S_1 \rightarrow {}^1$TT crossing event, impulsive Raman reference measurements on ^{1}TT were carried out after the initial singlet fission dynamics were complete (9 ps after photoexcitation, with a Raman pulse tuned to the ^{1}TT excited-state absorption, see Appendix 1 for details) [71]. In Fig. 8.4 the resulting intrinsic ^{1}TT Raman spectrum (purple) and the transferred $S_1 \rightarrow {}^1$TT spectrum isolated above (blue) are compared. Both exhibit the same high-frequency signature bands (> 1200 cm^{-1}) with similar intensity ratios. In the lower-frequency region, however, we observe numerous modes (127, 793, 1126 and 1207 cm^{-1}) with strongly enhanced intensities in the $S_1 \rightarrow {}^1$TT Raman spectrum.

Studies of photoinduced internal conversion in β-carotene [75] and rhodopsin [76] suggest that vibrational modes which show similar frequencies and relative intensities in both Raman spectra (intrinsic and transferred) can be assigned to tuning modes of the underlying photochemical process. Such modes are required in a process mediated by a conical intersection or avoided crossing to yield electronic degeneracy between initial and final electronic states, but they take no active part

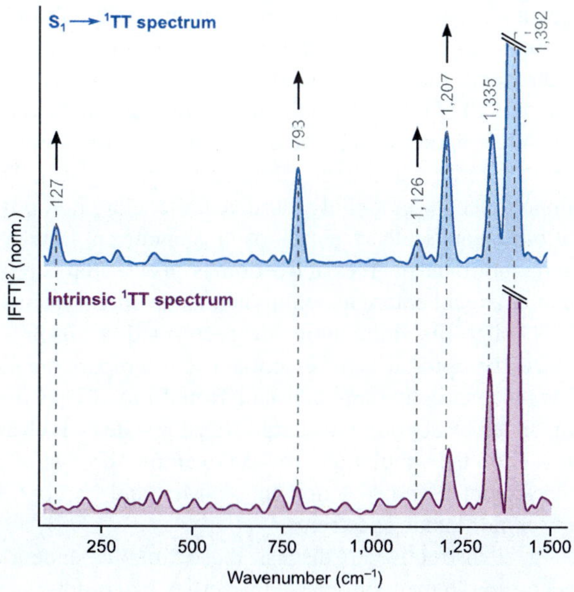

Fig. 8.4 Comparison of impulsive Raman spectra directly generated in ^{1}TT (bottom, purple) and transferred from S_1 via singlet fission (top, blue, reproduced from Fig. 8.3). Experimental spectra are normalised to the high-frequency peak at 1392 cm^{-1}, which is truncated for clarity. Arrows indicate regions where singlet fission increases the coherence intensity. Dashed vertical lines are a guide to the eye indicating the dominant vibrational modes

in the reaction [16]. In contrast, the presence of additional modes in the transferred coherence Raman spectrum has been largely unexplored. While the exact nature of these modes cannot be obtained from the experiments, the difference between transferred and intrinsically generated impulsive Raman spectra reflects a fundamental difference in the mechanism of vibrational coherence generation.

8.3.3 Simulation of Full Quantum Dynamics

To gain structural insight into this vibrationally coherent mechanism, we modelled the full quantum dynamics of fission employing a recently developed TTNS approach [49] which accounts for 252 vibrational modes spanning 110–1680 cm^{-1} and their respective couplings to 5 excited electronic states (see Sect. 8.2). This method expands the full vibronic wavefunction of the system into a network of tensors which represent molecular vibrations of different symmetries as well as the electronic system. All groups of vibrational tensors are connected (i.e. coupled) to the tensor representing the electronic system (Fig. 8.5a, left). This wavefunction is evolved in time using the time-dependent variational principle [77] in combination with a linear vibronic Hamiltonian, parametrised from *ab initio* calculations of the electronic and vibrational structure of DP-Mes. Crucially, not all correlations between the elements of the system are required to correctly capture its underlying ultrafast dynamics [49]. Upon re-expanding the wavefunction into a tree structure by means of entanglement renormalisation (Fig. 8.5a, right), we can include only the most significant correlations, making calculation of the time evolution computationally feasible even for large systems such as DP-Mes [52, 78]. This approach allows numerically exact quantum-mechanical treatment of many-body vibronic wavefunctions, without any recourse to perturbation theory or loss of information, as in reduced density matrix approaches that invoke Markov-like approximations. We note that despite the fact that not all of the 252 modes are strongly excited over the course of the dynamics, it remains important to include them in order to capture the effects of dissipation and irreversibility that a large vibrational bath imposes. This capability is of paramount importance for modelling ultrafast molecular photophysics—as well as a wide range of other nanoscale systems—as the concurrent time evolution of both the electronic and vibrational degrees of freedom is an essential part of the non-Born–Oppenheimer and emergent entanglement dynamics in such systems.

Though the simulations are performed at absolute zero, this theoretical model correctly reproduces the essential room-temperature electronic photophysics of DP-Mes—quantitative singlet fission from S_1 to ^1TT, as seen in Fig. 8.6. Singlet fission is mediated by coupling to charge-transfer states which are not directly populated [49]. Based on the simulations we can examine the role of each vibrational mode during the singlet fission reaction in DP-Mes and its contribution to the experimentally observed Raman spectrum.

It is instructive to discuss these time-dependent normal-mode displacements according to their symmetry properties by computing the total displacement ampli-

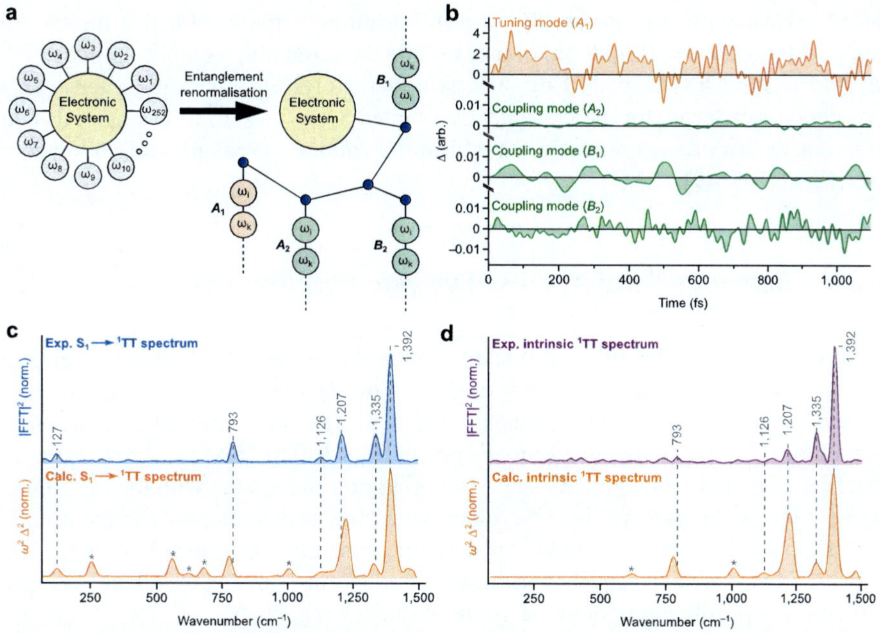

Fig. 8.5 **a** Schematic of TTNS. An electronic system is coupled to 252 vibrational modes (ω) in DP-Mes. By means of entanglement renormalisation, a tree of linearly connected vibrational modes connected to the electronic system via entanglement renormalisation nodes (blue) can be created, allowing facile computation of non-Born–Oppenheimer dynamics for large molecules (> 100 atoms). The vibrational modes are grouped by symmetry and colour coded to indicate tuning (orange) or coupling (green) mode behaviour. **b** Projected total time-dependent mode displacements for all symmetry groups and their corresponding assignment. We remark the $\approx$ 200-fold lower displacement amplitude for A_2, B_1 and B_2 modes compared to the A_1 modes. **c** $S_1 \rightarrow {}^1$TT coherence transfer Raman spectrum (blue) and resonance Raman representation of the calculated spectrum (orange). **d** Intrinsic ^{1}TT Raman spectrum (purple) compared to the resonance Raman representation of calculations initiated in ^{1}TT (orange). Calculated modes marked with an asterisk in **c** and **d** are not detected in the experimental spectrum, likely due to resonance Raman enhancement effects

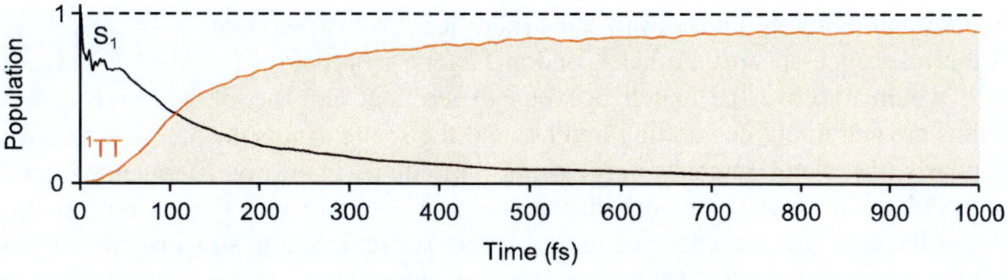

Fig. 8.6 Population dynamics of DP-Mes following photoexcitation of S_1. Charge transfer states are not directly populated and singlet fission towards the ^{1}TT state is quantitative

tude for each symmetry group (Fig. 8.5b). Intriguingly, the most active modes are the A_1 tuning modes (Fig. 8.5b, orange) while the remaining coupling modes (A_2, B_1 and B_2) are less displaced by at least a factor of 200 (Fig. 8.5b, green). This behaviour is expected for the vibronically coupled $S_1(B_2) \rightarrow TT(A_1)$ singlet fission reaction, as this process requires some form of symmetry-breaking motion to occur (see Fig. 8.1b) [49].

8.3.4 *Benchmarking Against Vibronic Spectroscopy*

To directly compare the theoretical results to the experimentally obtained coherence transfer Raman spectrum, we computed the Fourier transform of the total time-dependent normal-mode displacements and converted the retrieved displacement amplitudes (Δ) into resonance Raman spectra from 100 to 1500 cm^{-1} according to Eq. (8.6) (Fig. 8.5c, orange). We observe remarkable agreement with the coherence transfer Raman spectrum (Fig. 8.5c, blue), with clear activity not only in the prominent 1335 and 1392 cm^{-1} modes but also reproducing the spectrum at 127, 793, 1126 and 1207 cm^{-1} with a frequency accuracy of < 20 cm^{-1} and an excellent match to the intensity profile throughout the entire frequency region.

Our calculation also predicts several other strongly displaced modes in ^{1}TT, in particular in the region 250–800 cm^{-1}, which are not observed experimentally (asterisks in Fig. 8.5c). Whereas the simulation describes the full Raman spectrum of vibrational coherence generated after photoexcitation and transferred through singlet fission, this coherence can only be probed via the excited-state absorption transition at $\approx$ 520 nm. If a vibrational mode is not displaced along this (probed) transition, the experiment cannot benefit from the associated resonance Raman enhancement and will not display significant Raman intensity. Such a mode-specific resonance Raman effect is also commonly observed in linear resonance Raman spectroscopy upon tuning the excitation wavelength into different electronic transitions [72]. We therefore believe that the modes absent in the experimental spectrum in the region from 250 to 800 cm^{-1} are not Raman active on the $T_1 \rightarrow T_3$ absorption transition and are consequently invisible to the experiment. We expect that exact modelling of the resonance-specific Franck-Condon factors, as reported for TIPS-pentacene [79], would improve the match between experiment and theory, but such calculations are extremely demanding and beyond the scope of this chapter. Instead, we emphasise the overall intensity agreements, particularly in the high-frequency region ($>$ 1000 cm^{-1}). Such agreement, despite the fact our resonance Raman representation of the calculated displacements is only an approximation, supports the notion that the observed intensity differences between experiment and theory are related to mode-specific resonance Raman effects.

Two other potential explanations for these differences can be considered. Firstly, it has previously been reported that the singlet fission dynamics of DP-Mes are highly dependent on the environment [47, 67], a parameter not incorporated in the simulations. We anticipate that any change in the underlying mechanism will further affect the relative intensities of the observed Raman modes, complicating absolute intensity comparison between experiment (thin film) and theory (vacuum). It is encouraging in this regard that the experimental spectrum does not contain modes which were not predicted by the simulation. Secondly, we can consider that the additional modes (Fig. 8.5, starred) in the simulation arise coincidentally from the calculation. Since these modes cannot be detected in the experiment, we cannot directly confirm that they correspond to real molecular displacements. However, examining the absolute displacement parameters, we find that the low-frequency modes are among the most displaced in the entire system and therefore largely responsible for the overall reaction dynamics. Consequently, we would not expect our model to nearly quantitatively reproduce the timescale of singlet fission, the transfer of vibrational coherence and even the sensitivity of the ^{1}TT vibrational coherence to the way it is generated (see below), if these modes were spurious in origin. This accuracy, benchmarked on multiple observables, suggests that the underlying linear vibronic Hamiltonian provides a satisfactory description of DP-Mes, and that the additional modes in the calculation are not coincidental but more likely absent in the experiment due to Raman enhancement and environmental factors.

To further validate the structural sensitivity of our theoretical framework, we performed the same analysis for a trajectory initiated in the ^{1}TT state, generating its intrinsic vibrational structure. In analogy to the coherence transfer spectrum, we find that the dominantly displaced modes belong again to A_1 symmetry, with negligible contributions of other modes. The resonance Raman spectrum resulting from this intrinsic calculation (Fig. 8.5d, orange) again predicts some modes which are not experimentally observed (see above), but importantly reproduces every experimental ^{1}TT frequency. Comparing the relative intensities between the experimental and theoretical spectrum is subject to the same resonance Raman considerations mentioned above. Importantly, the more complex nature of the intrinsic ^{1}TT Raman experiment is expected to result in larger deviations in the relative mode intensities compared to the calculation due to multiple resonance Raman enhancement effects.

Our simulations further allow us to compare directly the effect of initiating the trajectory on S_1 or ^{1}TT without the added complication of varying resonance Raman factors. In the high-frequency region ($> 1000\,\mathrm{cm}^{-1}$), we observe a near-perfect intensity match with marginal intensity differences only noticeable for the 1207 cm^{-1} mode. Crucially, in the low-frequency region, the simulations reproduce the surprising observation of a new low-frequency mode at 127 cm^{-1} that only appears following coherence transfer from S_1 (Fig. 8.4), as a signature of singlet fission. Despite any ambiguities regarding optical selection rules or the role of environment, comparison between S_1-initiated and ^{1}TT-initiated vibrational coherence reveals the same essential results in experiment and theory: singlet fission in DP-Mes enables transfer of vibrational coherence, and this coherence carries unique signatures of the passage of the wave-packet. Thus we consider that our simulations provide an excellent description of the tuning modes in DP-Mes. We further recall that the simulations closely reproduce the electronic dynamics [49], including the mediating role of virtual charge-transfer states [47, 67]. As these dynamics depend sensitively on the interplay of tuning and coupling modes, our model description of the coupling modes is at a minimum qualitatively correct.

8.3.5 Character of Coupling and Tuning Modes

Motivated by the striking experimental and theoretical structural agreement, we can assign the specific role of each observed mode governing the singlet fission process in DP-Mes. Figures 8.5b, c emphasise that all experimentally observed modes in the coherence transfer spectrum are of A_1 symmetry, and consequently do not mediate the coupling between S_1 and ^1TT. Displacements of these modes alter the energies of the states but the vibronic couplings between them remain zero. The experimental coherence transfer spectrum therefore represents the most displaced tuning modes of the singlet fission reaction. From our *ab initio* simulations, we can immediately show that their atomistic vibrational motions are primarily associated with in-plane ring deformations affecting each pentacene moiety as well as modulating the pentacene-pentacene central bond lengths (Fig. 8.7a, left). This is consistent with the behaviour expected for tuning modes in conjugated systems where bond-length alterations are caused by the optically induced $\pi \rightarrow \pi^*$ transition to energetically relax the molecule [9, 80, 81].

According to our TTNS simulations the most active coupling modes of the reaction are instead of B_1 and B_2 symmetry, albeit with predicted displacements that are at least two orders of magnitude smaller than the tuning modes (Fig. 8.5b). In line with this result, the experimental resonance Raman spectra reveal no peaks directly attributable to coupling modes. The quadratic dependence of the Raman intensity on the mode displacements (Eq. 8.6) implies these would be weaker in intensity by at least 4 orders of magnitude, which is too small to be detected even with a high signal-to-noise ratio. This observation, specifically illustrated here for DP-Mes, likely explains the general lack of experimentally observed coupling modes in singlet fission and similar condensed-phase surface crossing reactions in the literature.

While here it was not possible to experimentally monitor the coupling modes, preventing us from directly validating their simulation parameters, we emphasise that our simulations accurately describe the singlet fission dynamics as well as the transfer of vibrational coherence (including the unique enhancement of low-frequency modes). These observables are governed by the precise interplay of both tuning and coupling modes, such that, if the coupling mode description in our simulations were inadequate, the model would fail to reproduce the overall dynamics and coherence transfer characteristics. We carried out additional simulations by systematically varying the coupling mode strength, which confirm the notion that these degrees of freedom are strongly linked and that the coupling mode description is well constrained. An in-depth discussion of these results is provided in Appendix 3.

A fundamental limitation in theoretically describing a fully structural model of electronic dynamics is that the model contains more parameters than observables. Our approach benchmarks the model against not just the timescale of singlet fission but also the spectrum of ^1TT vibrational coherence and its sensitivity to the fission pathway, allowing for a greater degree of confidence in the theoretical description than previously achieved. Despite potential for an improved model description, we are confident to extract the main structural character of the coupling modes from our TTNS simulations. We find that the primary effect of the coupling

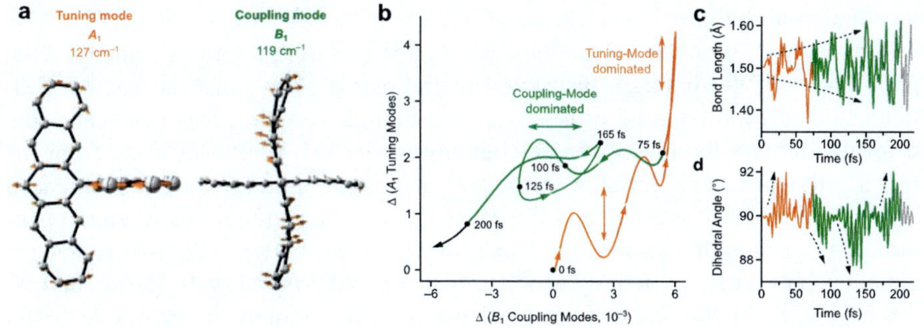

Fig. 8.7 **a** Representative normal mode examples of a tuning (left) and a coupling mode (right). The molecular structure was truncated to the core pentacene units for clarity. **b** Correlation plot of the total time-dependent displacement amplitude of A_1 tuning and B_1 coupling modes. Contour arrows indicate the direction of evolution and black dots indicate the time of the trajectory. The initial motion is dominated by tuning modes (orange vertical arrows) before a rapid damping of the tuning modes funnels the energy into the coupling modes (green horizontal arrow). **c** Central pentacene-pentacene bond length and **d** dihedral angle during the ultrafast singlet fission reaction. Traces are colour coded to match panel **b** and dashed arrows indicate significant increases in the values of the parameters

modes is to create a local twist around the pentacene-pentacene bond and thus a deviation from orthogonality (see Fig. 8.7a, right). This localised twist causes transient, time-dependent wavefunction overlap of the frontier orbitals of each pentacene monomer. The resulting electronic coupling between the pentacene sub-units thereby promotes the superexchange reaction from S_1 to ^{1}TT via the high-lying CT states [24, 47, 49, 67].

8.3.6 Coordinated Interplay of Coupling and Tuning Modes

To understand the interplay of vibrational tuning and coupling coordinates during and following the initial Franck-Condon relaxation ($\approx$ 200 fs), we turn to the detailed molecular movie generated in our simulations (please refer to the supplementary material of [1] for the full movie). Upon investigating the correlation between the collective time-dependent displacements of the A_1 (tuning) and B_1 (coupling) modes within the first 200 fs (Fig. 8.7b) we identify two sequential temporal regimes. From 0 to 75 fs, the dynamics are dominated by tuning modes (vertical displacement, orange), which position the energy levels of DP-Mes for efficient $S_1 \rightarrow {}^1$TT crossing. Over the subsequent 125 fs, the activity shifts towards the coupling modes (horizontal displacement, green) which drive the conversion. Importantly, during this coupling-mode dominated regime, the tuning mode displacements are severely damped to further enhance the reaction yield, highlighting that the singlet fission reaction in DP-Mes is dictated by the synchronised motions along these collective coordinates.

To explore the effect of this collective motion on local coordinates, we extracted from our movie the time-evolution of two key structure parameters—the central pentacene dimer bond length and associated dihedral angle. We find that the amplitude modulation of the inter-pentacene bond steadily increases until $\approx$ 150 fs (Fig. 8.7c, dashed arrows), with the largest increase occurring during the tuning-mode driven period at $\approx$ 65 fs (orange). Inspection of the movie shows that the local activation of this central bond is furthermore coupled to several in-plane pentacene ring deformations. As the much weaker coupling modes drive DP-Mes away from the initial orthogonal configuration of the two pentacene units (green), this enhanced bond length activity substantially increases the coupling strength between the pentacene π-systems. In contrast, the local dihedral angle between the two pentacene units displays bursts of activity at well-separated $\approx$ 50 fs intervals during the first 200 fs (Fig. 8.7c). We find that positive bursts are present in the tuning-mode driven time window (orange), while negative bursts correlate with the coupling-mode driven regime (green and dashed arrows). Such local behaviour depicts a well-defined evolution on the potential energy surfaces during which the molecule is regularly returned to an energetically favourable region where coupling modes can actively drive the reaction. We note that while these motions are initially activated through the singlet fission process, they do not simply track the electronic kinetics but remain active throughout the simulation window.

8.4 Conclusions

Our approach provides clarity about the ultrafast intramolecular singlet fission reaction in DP-Mes with significantly improved structural resolution. Structurally sensitive excited-state Raman spectroscopy was employed to uncover the transfer of vibrational wavepackets from S_1 to ^1TT, mandating a vibrationally coherent reaction mechanism despite there being no direct coupling between these states. Using this detailed kinetic and structural information as a benchmark for a full quantum dynamics simulation enabled us to assign the observed Raman spectrum to the dominant tuning modes of the process. The compelling match between simulation and experiment allowed us further to infer the crucial coupling modes of the system and record the underlying molecular movie for the singlet fission reaction in a complex molecular system of over 100 atoms. While we have focused exclusively on intramolecular singlet fission in this chapter, both the experimental and theoretical techniques can be equally well applied to a wide range of ultrafast photochemical processes, from charge transfer [2] and photoisomerisation [9] to polaritonic chemistry [82, 83]. Together, these methods set up a powerful tool for describing and understanding reaction dynamics beyond the Born–Oppenheimer approximation, providing detailed insight into the reaction mechanism at conical intersections or avoided crossings in complex materials of practical interest. One immediate result from our analysis is that the coupling modes underpinning singlet fission in DP-Mes are far less intense than the tuning modes. In this and likely most other systems, new experimental approaches are needed to directly observe coupling modes. Alternatively, the TTNS simulations could be systematically improved to explore the coupling mode parameter spaces within the linear vibronic Hamiltonian description—the quality of which underpins the full simulation. The robustness of the comparison to experiment could be further enhanced by direct incorporation of finite-temperature and environmental effects within the TTNS method , or incorporation of the relevant Franck-Condon factors [79] in the transformation into resonance Raman spectra.

Beyond establishing detection limits for mechanistically relevant vibrational modes, our results report on the functionality of coupled vibrational and electronic dynamics in ultrafast reactions. The photoexcited system evolves along a precisely synchronised set of multiple vibrational modes which must act in concert (i.e. coherently) to promote a highly efficient ultrafast reaction. Taken together with recent results on a vibrational-phase effect on the ultrafast photoisomerisation reaction in rhodopsin [9], our results strongly support the notion that the vibrationally coherent evolution out of the Franck-Condon region after photoexcitation offers an opportunity to tune the outcome of any ultrafast (< 10 ps) reaction through rational design [3]. Indeed, the ability to visualise molecular motion with atomistic detail offers prediction scope for systems that exhibit a strong connection between functionality and real-space motion, paving the way for the discovery of novel functional materials. We anticipate the techniques presented above will enable significant advances in the understanding of ultrafast phenomena in general, from charge generation in solar cells [84] to biological light harvesting [85].

Appendix 1: Experimental Details

While the experimental work presented in this chapter was not undertaken by the author of this thesis, the relevant details are given here for completeness.

1.1 Impulsive Vibrational Spectroscopy Setup

The impulsive vibrational spectroscopy setup has been described in detail elsewhere [12, 71]. Briefly, a Yb:KGW amplifier system (Light Conversion, Pharos, 5 W, 10 kHz) provides pulses centred at 1030 nm with a pulse duration of $\approx$ 200 fs. A small portion is used to generate a chirped white light continuum in a 3 mm sapphire window used as the probe pulse in all experiments, with a Gaussian beam diameter at full-width-half-maximum (fwhm) of 30 μm at the sample. Impulsive pump pulses in the near-IR (150 nJ, 50 μm fwhm) and visible (130 nJ, 70 μm fwhm) were generated by previously reported home-built non-collinear parametric amplifiers (NOPAs) [86]. To produce the narrow-band pump pulse (80 nJ, 70 μm fwhm) used for the preparation of the triplet state (Fig. 8.4) the seed white light continuum was temporally stretched in the visible pump NOPA with a BK7 rod (7 cm length) prior to amplification resulting in $\approx$ 200 fs pulses. The duration is limited by the employed pump pulses (third harmonic) which was derived directly from the amplifier system.

1.2 Non-resonant Impulsive Vibrational Spectroscopy

To gain access to the ground-state Raman spectrum, a temporally compressed 12 fs pump pulse tuned to 800 nm was employed, which generates vibrational coherences via impulsive stimulated Raman scattering exclusively on the ground electronic state due to the lack of an electronic resonance (see Fig. 8.4c, right). Fourier transformation of the detected coherent oscillations over the absorption spectrum of DP-Mes makes it possible to independently measure a time-domain Raman spectrum which is directly comparable to resonant impulsive Raman spectra when probed in the same wavelength region [69, 71, 73].

1.3 Excited-State Impulsive Vibrational Spectroscopy

Intrinsic impulsive Raman reference measurements on ^{1}TT were carried out by first photoexciting the sample (600 nm, 200 fs) to generate a population in S_1 that was allowed to undergo singlet fission. After a time delay of 9 ps, at which point the singlet fission reaction is complete [47], the ^{1}TT population was re-excited with an impulsive Raman pump pulse (800 nm, 11 fs) resonant with the excited-state absorption observed $>$ 750 nm [47]. The impulsively generated vibrational coherence was subsequently recorded in the excited-state absorption region (515–525 nm). Following subtraction of electronic kinetics and the background ground-state vibrational activity, a Fourier transform provided the intrinsic triplet Raman signature (Fig. 8.4, purple). No significant vibrational activity is detected in the ground-state bleach or stimulated emission regions after the subtraction procedure.

1.4 Vibrational Spectroscopy Data Analysis

The experimental impulsive vibrational spectroscopy datasets were processed according to previously published procedures [71]. Briefly, after chirp-correction, all traces were truncated in time to include only positive time delays $>$ 210 fs to prevent coherent artefact contributions affecting the signal. The residual oscillations for each probe wavelength were subsequently extracted by globally fitting the experimentally recorded maps to a sum of two exponentially decaying functions with an offset. The coherent oscillations were further truncated to an overall time length of 1.28 ps prior to apodisation (Kaiser–Bessel window, $\beta = 1$), zero-padding (3$\times$) and Fourier transformation. The frequency resolution corresponded to ≈ 26 cm^{-1} and the lowest resolvable frequency was ≈ 52 cm^{-1}. To extract the intrinsic ^{1}TT Raman spectrum, the vibrational coherence in the presence and absence of the actinic pump pulse was recorded and subsequently the two coherences in the time-domain were subtracted to minimise ground-state contributions, as outlined previously. The remaining data analysis was the same. Care was taken to ensure that all traces were temporally aligned using reference measurements on toluene to exclude possible Fourier artefacts in the comparison between different experiments.

Appendix 2: Vibronic Coupling Matrices

Below are the W matrices appearing in the linear vibronic Hamiltonian of Eq. (8.1).

$$\overline{W}_{A_{1,1}} = -\begin{bmatrix} 0 & 0 & 0 & 0 & 0 \\ 0 & 0.45 & 0 & 0 & 0 \\ 0 & 0 & 0.38 & 0 & 0 \\ 0 & 0 & 0 & 0.57 & 0 \\ 0 & 0 & 0 & 0 & 0.57 \end{bmatrix},$$
$$\overline{W}_{A_{1,2}} = -\begin{bmatrix} 0.75 & 0 & 0 & 0 & 0 \\ 0 & 0.3 & 0 & 0 & 0 \\ 0 & 0 & 0.29 & 0 & 0 \\ 0 & 0 & 0 & 0.37 & 0 \\ 0 & 0 & 0 & 0 & 0.37 \end{bmatrix},$$
$$\overline{W}_{A_2} = -\begin{bmatrix} 0 & 0 & 0 & 1 & 0 \\ 0 & 0 & 0 & 0 & 0 \\ 0 & 0 & 0 & 0 & 0 \\ 1 & 0 & 0 & 0 & 0 \\ 0 & 0 & 0 & 0 & 0 \end{bmatrix},$$
$$\overline{W}_{B_{1,1}} = -\begin{bmatrix} 0 & 0 & 0 & 0 & -0.63 \\ 0 & 0 & 0 & 0.73 & 0 \\ 0 & 0 & 0 & 0 & -0.26 \\ 0 & 0.73 & 0 & 0 & 0 \\ -0.63 & 0 & -0.26 & 0 & 0 \end{bmatrix}, \tag{8.7}$$
$$\overline{W}_{B_{1,2}} = -\begin{bmatrix} 0 & 0 & 0 & 0 & 0.28 \\ 0 & 0 & 0 & -0.33 & 0 \\ 0 & 0 & 0 & 0 & -0.9 \\ 0 & -0.33 & 0 & 0 & 0 \\ 0.28 & 0 & -0.9 & 0 & 0 \end{bmatrix},$$
$$\overline{W}_{B_{2,1}} = -\begin{bmatrix} 0 & 0 & 0 & 0 & 0 \\ 0 & 0 & 0.86 & 0 & 0 \\ 0 & 0.86 & 0 & 0 & 0 \\ 0 & 0 & 0 & 0 & 0.52 \\ 0 & 0 & 0 & 0.52 & 0 \end{bmatrix},$$
$$\overline{W}_{B_{2,2}} = -\begin{bmatrix} 0 & 0 & 0 & 0 & 0 \\ 0 & 0 & -0.86 & 0 & 0 \\ 0 & -0.86 & 0 & 0 & 0 \\ 0 & 0 & 0 & 0 & 0.52 \\ 0 & 0 & 0 & 0.52 & 0 \end{bmatrix}.$$

Modes of A_1 symmetry are tuning modes, while the rest of them couple different combinations of diabatic states to each other.

Appendix 3: Coupling Mode Assignment

Our theoretical approach predicts the real-time displacement of all the vibrational modes of the system. However, it is only the tuning modes which can directly be compared to the experimental resonant Raman spectra, as the displacement of coupling modes is orders of magnitude smaller and cannot be detected. Despite the fact that for the tuning modes we find an excellent agreement with experiment, it is necessary to examine whether the results for the coupling modes are equally valid. To this end, we alter the coupling mode constants universally, by introducing an additional scaling parameter γ_i in the linear vibronic Hamiltonian:

$$H_{star} = H_{el} + \sum_i \gamma_i \cdot \overline{W_i} \sum_k \lambda_{i,k} \frac{b^\dagger_{i,k} + b_{i,k}}{\sqrt{2}} + H_{c,i}, \tag{8.8}$$

and we vary its value from 0.75 to 1.15 in steps of 0.05 for all clusters of coupling modes $i = \{A_2, B_{1,1}, B_{1,2}, B_{2,1}, B_{2,2}\}$. For the tuning modes we keep its value at one. We find that even reducing the coupling mode constants to 95% of their *ab initio* values slows singlet fission down by 37%. The full results are summarised in Fig. 8.8. Therefore, it becomes clear that coupling modes have a large impact on the overall singlet fission dynamics. The population dynamics which are obtained using the *ab initio* values of the coupling mode constants are found to be in good agreement with the experimental singlet fission kinetics, indicating that the properties of coupling modes which are used in our model must be at least semi-quantitative.

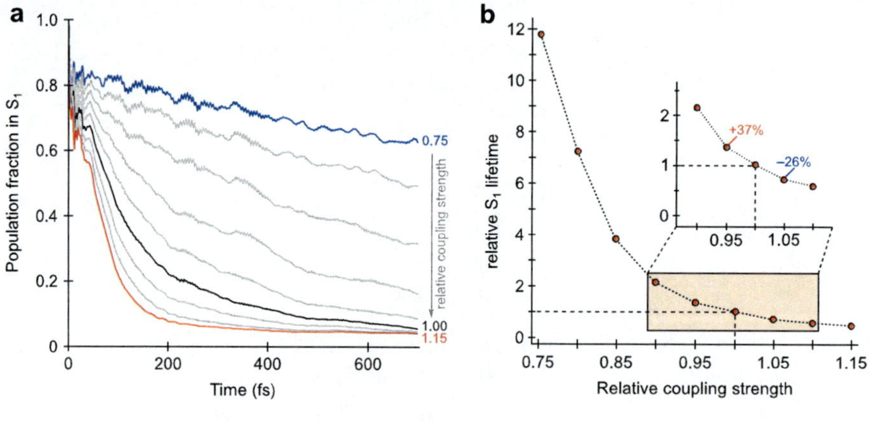

Fig. 8.8 **a** S_1 decay dynamics resulting from modifying the coupling mode strengths in the range from 0.75 to 1.15. **b** Retrieved ratio of single-exponential time constants relative to the native (1.00) coupling strength of the singlet decay dynamics. Inset shows a close-up in the region around 1.00 for clarity

Appendix 4: Comparing Theory and Experiment

TTNS provides time-dependent displacements of all vibrational modes grouped by their respective symmetry groups. After projection onto the DFT-calculated normal modes of DP-Mes, we constructed the total time-dependent displacements for each symmetry group (Fig. 8.5b). We subsequently selected only the A_1 mode displacements due to their drastically enhanced activity and applied the same data analysis as for the experimental traces, i.e. apodisation (Kaiser-Bessel window, $\beta = 1$), zero-padding ($3\times$) and Fourier transformation. The simulated time-vector was marginally shorter (1.12 ps), leading to a frequency resolution of $\approx$ 29 cm^{-1} with a lowest resolvable frequency of $\approx$ 59 cm^{-1}. After Fourier transformation, the displacement spectrum was scaled in frequency by 0.97 before computing Eq. (8.6). This scaling factor was determined by a matching the DFT-calculated Raman active modes to the ground-state Raman spectrum of DP-Mes. The same approach was carried out for the simulations that were initiated in ^{1}TT.

References

1. Schnedermann C, Alvertis AM, Wende T, Lukman S, Feng J, Schröder FAYN, Turban DHP, Wu J, Hine NDM, Greenham NC, Chin AW, Rao A, Kukura P, Musser AJ (2019) A molecular movie of ultrafast singlet fission. Nat Commun 10:4207
2. Falke SM, Rozzi CA, Brida D, Maluri M, Amato M, Sommer E, De Sio A, Rubio A, Cerullo G, Molinari E, Lienau C (2014) Coherent ultrafast charge transfer in an organic photovoltaic blend. Science 344:1001
3. Delor M, Archer SA, Keane T, Meijer AJHM, Sazanovich IV, Greetham GM, Towrie M, Weinstein JA (2017) Directing the path of light-induced electron transfer at a molecular fork using vibrational excitation. Nat Chem 9:1099
4. Bakulin AA, Lovrincic R, Yu X, Selig O, Bakker HJ, Rezus YLA, Nayak PK, Fonari A, Coropceanu V, Brédas JL, Cahen D (2015) Mode-selective vibrational modulation of charge transport in organic electronic devices. Nat Commun 6:7880
5. Sukegawa J, Schubert C, Zhu X, Tsuji H, Guldi DM, Nakamura E (2014) Electron transfer through rigid organic molecular wires enhanced by electronic and electron-vibration coupling. Nat Chem 6:899
6. Delor M, Scattergood PA, Sazanovich IV, Parker AW, Greetham GM, Meijer AJHM, Towrie M, Weinstein JA (2014) Towards control of electron transfer in donor-acceptor molecules by bond-specific infrared excitation. Science 346:1492
7. Polli D, Altoè P, Weingart O, Spillane KM, Manzoni C, Brida D, Tomasello G, Orlandi G, Kukura P, Mathies RA, Garavelli M, Cerullo G (2010) Conical intersection dynamics of the primary photoisomerization event in vision. Nature 467:440
8. Roberts GM, Hadden DJ, Bergendahl LT, Wenge AM, Harris SJ, Karsili TNV, Ashfold MNR, Paterson MJ, Stavros VG (2013) Exploring quantum phenomena and vibrational control in σ^* mediated photochemistry. Chem Sci 4:993
9. Schnedermann C, Yang X, Liebel M, Spillane KM, Lugtenburg J, Fernández I, Valentini A, Schapiro I, Olivucci M, Kukura P, Mathies RA (2018) Evidence for a vibrational phase-dependent isotope effect on the photochemistry of vision. Nat Chem 10:449
10. Scholes GD, Fleming GR, Chen LX, Aspuru-Guzik A, Buchleitner A, Coker DF, Engel GS, Van Grondelle R, Ishizaki A, Jonas DM, Lundeen JS, McCusker JK, Mukamel S, Ogilvie JP, Olaya-Castro A, Ratner MA, Spano FC, Whaley KB, Zhu X (2017) Using coherence to enhance function in chemical and biophysical systems. Nature 543:647

11. Scholes GD, Fleming GR, Olaya-Castro A, Van Grondelle R (2011) Lessons from nature about solar light harvesting. Nat Chem 3:763
12. Musser AJ, Liebel M, Schnedermann C, Wende T, Kehoe TB, Rao A, Kukura P (2015) Evidence for conical intersection dynamics mediating ultrafast singlet exciton fission. Nat Phys 11:352
13. Bakulin AA, Morgan SE, Kehoe TB, Wilson MWB, Chin AW, Zigmantas D, Egorova D, Rao A (2016) Real-time observation of multiexcitonic states in ultrafast singlet fission using coherent 2D electronic spectroscopy. Nat Chem 8:16
14. Butler LJ (1998) Chemical reaction Born-Oppenheimer approximation. Annu Rev Phys Chem 49:125
15. Balzer B, Hahn S, Stock G (2003) Mechanism of a photochemical funnel: a dissipative wave-packet dynamics study. Chem Phys Lett 379:351
16. Kühl A, Domcke W (2002) Multilevel Redfield description of the dissipative dynamics at conical intersections. J Chem Phys 116:263
17. Egorova D, Domcke W (2004) Quantum dynamical simulations of ultrafast photoinduced electron-transfer processes. J Photochem Photobiol A Chem 166:19
18. Rafiq S, Scholes GD (2019) From fundamental theories to quantum coherences in electron transfer. J Am Chem Soc 141:708
19. Smith MB, Michl J (2010) Singlet fission. Chem Rev 110:6891
20. Wilson MWB, Rao A, Clark J, Kumar RSS, Brida D, Cerullo G, Friend RH (2011) Ultrafast dynamics of exciton fission in polycrystalline pentacene. J Am Chem Soc 133:11830
21. Chan WL, Ligges M, Jailaubekov AE, Kaake LG, Miaja-Avila L, Zhu X-Y (2011) Observing the multiexciton state in singlet fission and ensuing ultrafast multielectron transfer. Science 334:1541
22. Yost SR, Lee J, Wilson MWB, Wu T, McMahon DP, Parkhurst RR, Thompson NJ, Congreve DN, Rao A, Johnson K, Sfeir MY, Bawendi MG, Swager TM, Friend RH, Baldo M, Van Voorhis T (2014) A transferable model for singlet-fission kinetics. Nat Chem 6:492
23. Zimmerman PM, Zhang Z, Musgrave CB (2010) Singlet fission in pentacene through multi-exciton quantum states. Nat Chem 2:648
24. Berkelbach TC, Hybertsen MS, Reichman DR (2013) Microscopic theory of singlet exciton fission. I. General formulation. J Chem Phys 138:114102
25. Stern HL, Cheminal A, Yost SR, Broch K, Bayliss SL, Chen K, Tabachnyk M, Thorley K, Greenham N, Hodgkiss JM, Anthony J, Head-Gordon M, Musser AJ, Rao A, Friend RH (2017) Vibronically coherent ultrafast triplet-pair formation and subsequent thermally activated dissociation control efficient endothermic singlet fission. Nat Chem 9:1205
26. Miyata K, Kurashige Y, Watanabe K, Sugimoto T, Takahashi S, Tanaka S, Takeya J, Yanai T, Matsumoto Y (2017) Coherent singlet fission activated by symmetry breaking. Nat Chem 9:983
27. Monahan NR, Sun D, Tamura H, Williams KW, Xu B, Zhong Y, Kumar B, Nuckolls C, Harutyunyan AR, Chen G, Dai HL, Beljonne D, Rao Y, Zhu XY (2017) Dynamics of the triplet-pair state reveals the likely coexistence of coherent and incoherent singlet fission in crystalline hexacene. Nat Chem 9:341
28. Breen I, Tempelaar R, Bizimana LA, Kloss B, Reichman DR, Turner DB (2017) Triplet separation drives singlet fission after femtosecond correlated triplet pair production in rubrene. J Am Chem Soc 139:11745
29. Bera K, Douglas CJ, Frontiera RR (2017) Femtosecond Raman microscopy reveals structural dynamics leading to triplet separation in rubrene singlet fission. J Phys Chem Lett 8:5929
30. Tempelaar R, Reichman DR (2018) Vibronic exciton theory of singlet fission. III. How vibronic coupling and thermodynamics promote rapid triplet generation in pentacene crystals. J Chem Phys 148:244701
31. Fujihashi Y, Chen L, Ishizaki A, Wang J, Zhao Y (2017) Effect of high-frequency modes on singlet fission dynamics. J Chem Phys 146:044101
32. Monserrat B, Engel EA, Needs RJ (2015) Giant electron-phonon interactions in molecular crystals and the importance of nonquadratic coupling. Phys Rev B 92:140302(R)

33. Zheng YJ, Xie SJ, Lan Z (2016) Ultrafast nonadiabatic dynamics of singlet fission: quantum dynamics with the multilayer multiconfigurational time-dependent Hartree (ML-MCTDH) method. J Phys Chem C 120:1375
34. Teichen PE, Eaves JD (2012) A microscopic model of singlet fission. J Phys Chem B 116:11473
35. Renaud N, Grozema FC (2015) Intermolecular vibrational modes speed up singlet fission in perylenediimide crystals. J Phys Chem Lett 6:360
36. Reddy SR, Coto PB, Thoss M (2018) Intramolecular singlet fission: insights from quantum dynamical simulations. J Phys Chem Lett 9:5979
37. Tamura H, Huix-Rotllant M, Burghardt I, Olivier Y, Beljonne D (2015) First-principles quantum dynamics of singlet fission: coherent versus thermally activated mechanisms governed by molecular π stacking. Phys Rev Lett 115:107401
38. Feng X, Luzanov AV, Krylov AI (2013) Fission of entangled spins: an electronic structure perspective. J Phys Chem Lett 4:3845
39. Zeng T, Hoffmann R, Ananth N (2014) The low-lying electronic states of pentacene and their roles in singlet fission. J Am Chem Soc 136:5755
40. Castellanos MA, Huo P (2017) Enhancing singlet fission dynamics by suppressing destructive interference between charge-transfer pathways. J Phys Chem Lett 8:2480
41. Feng X, Krylov AI (2016) On couplings and excimers: lessons from studies of singlet fission in covalently linked tetracene dimers. Phys Chem Chem Phys 18:7751
42. Huang Z, Fujihashi Y, Zhao Y (2017) Effect of off-diagonal exciton-phonon coupling on intramolecular singlet fission. J Phys Chem Lett 8:3306
43. Casanova D (2018) Theoretical modeling of singlet fission. Chem Rev 118:7164
44. Zirzlmeier J, Lehnherr D, Coto PB, Chernick ET, Casillas R, Basel BS, Thoss M, Tykwinski RR, Guldi DM (2015) Singlet fission in pentacene dimers. Proc Natl Acad Sci 112:5325
45. Sanders SN, Kumarasamy E, Pun AB, Trinh MT, Choi B, Xia J, Taffet EJ, Low JZ, Miller JR, Roy X, Zhu XY, Steigerwald ML, Sfeir MY, Campos LM (2015) Quantitative intramolecular singlet fission in bipentacenes. J Am Chem Soc 137:8965
46. Kumarasamy E, Sanders SN, Tayebjee MJY, Asadpoordarvish A, Hele TJH, Fuemmeler EG, Pun AB, Yablon LM, Low JZ, Paley DW, Dean JC, Choi B, Scholes GD, Steigerwald ML, Ananth N, McCamey DR, Sfeir MY, Campos LM (2017) Tuning singlet fission in π-bridge-π chromophores. J Am Chem Soc 139:12488
47. Lukman S, Chen K, Hodgkiss JM, Turban DHP, Hine NDM, Dong S, Wu J, Greenham NC, Musser AJ (2016) Tuning the role of charge-transfer states in intramolecular singlet exciton fission through side-group engineering. Nat Commun 7:13622
48. Lukman S, Richter JM, Yang L, Pan H, Jishan W, Greenham NC, Musser AJ (2017) Efficient singlet fission and triplet-pair emission in a family of zethrene diradicaloids. J Am Chem Soc 139:18376
49. Schröder FAYN, Turban DHP, Musser AJ, Hine NDM, Chin AW (2019) Tensor network simulation of multi-environmental open quantum dynamics via machine learning and entanglement renormalisation. Nat Commun 10:1062
50. Schollwöck U (2011) The density-matrix renormalization group in the age of matrix product states. Ann Phys 326:96
51. Orús R (2014) A practical introduction to tensor networks: matrix product states and projected entangled pair states. Ann Phys 349:117
52. Vidal G (2007) Entanglement renormalization. Phys Rev Lett 99:220405
53. Hackbusch W, Khoromskij BN, Tyrtyshnikov EE (2005) Hierarchical Kronecker tensor-product approximations. J Numer Math 13:119
54. Hackbusch W (2012) Tensor spaces and numerical tensor calculus. Springer, Berlin Heidelberg
55. Lubich C, Rohwedder T, Schneider R, Vandereycken B (2013) Dynamic approximation by hierarchical tucker and tensor-train tensors. SIAM J Matrix Anal Appl 34:470
56. Chin AW, Rivas Á, Huelga SF, Plenio MB (2010) Exact mapping between system-reservoir quantum models and semi-infinite discrete chains using orthogonal polynomials. J Math Phys 51:092109

57. Chin AW, Prior J, Rosenbach R, Caycedo-Soler F, Huelga SF, Plenio MB (2013) The role of non-equilibrium vibrational structures in electronic coherence and recoherence in pigment-protein complexes. Nat Phys 9:113
58. Guo C, Weichselbaum A, von Delft J, Vojta M (2012) Critical and strong-coupling phases in one- and two-bath spin-boson models. Phys Rev Lett 108:160401
59. Prior J, Chin AW, Huelga SF, Plenio MB (2010) Efficient simulation of strong system-environment interactions. Phys Rev Lett 105:050404
60. Schröder FAYN, Chin AW (2016) Simulating open quantum dynamics with time-dependent variational matrix product states: towards microscopic correlation of environment dynamics and reduced system evolution. Phys Rev B 93:075105
61. Gonzalez-Ballestero C, Schröder FAYN, Chin AW (2017) Uncovering nonperturbative dynamics of the biased sub-Ohmic spin-boson model with variational matrix product states. Phys Rev B 96:115427
62. Wall ML, Safavi-Naini A, Rey AM (2016) Simulating generic spin-boson models with matrix product states. Phys Rev A 94:053637
63. Okada K, Tonami T, Nagami T, Nakano M (2019) Breakdown of the perturbative approach to molecular packing dependence of singlet fission rates in pentacene dimer models: a systematic comparison with the quantum master equation approach. J Phys Chem C 123:15403
64. Kollmar C (1993) Electronic structure of diradical and dicarbene intermediates in short-chain polydiacetylene oligomers. J Chem Phys 98:7210
65. Musser AJ, Clark J (2019) Triplet-pair states in organic semiconductors. Annu Rev Phys Chem 70:323
66. Aprà E, Bylaska EJ, de Jong WA, Govind N, Kowalski K, Straatsma TP, Valiev M, van Dam HJJ, Alexeev Y, Anchell J, Anisimov V, Aquino FW, Atta-Fynn R, Autschbach J, Bauman NP, Becca JC, Bernholdt DE, Bhaskaran-Nair K, Bogatko S, Borowski P, Boschen J, Brabec J, Bruner A, Cauët E, Chen Y, Chuev GN, Cramer CJ, Daily J, Deegan MJO, Dunning TH, Dupuis M, Dyall KG, Fann GI, Fischer SA, Fonari A, Früchtl H, Gagliardi L, Garza J, Gawande N, Ghosh S, Glaesemann K, Götz AW, Hammond J, Helms V, Hermes ED, Hirao K, Hirata S, Jacquelin M, Jensen L, Johnson BG, Jónsson H, Kendall RA, Klemm M, Kobayashi R, Konkov V, Krishnamoorthy S, Krishnan M, Lin Z, Lins RD, Littlefield RJ, Logsdail AJ, Lopata K, Ma W, Marenich AV, Martin Del Campo J, Mejia-Rodriguez D, Moore JE, Mullin JM, Nakajima T, Nascimento DR, Nichols JA, Nichols PJ, Nieplocha J, Otero-de-la Roza A, Palmer B, Panyala A, Pirojsirikul T, Peng B, Peverati R, Pittner J, Pollack L, Richard RM, Sadayappan P, Schatz GC, Shelton WA, Silverstein DW, Smith DMA, Soares TA, Song D, Swart M, Taylor HL, Thomas GS, Tipparaju V, Truhlar DG, Tsemekhman K, Van Voorhis T, Vázquez-Mayagoitia A, Verma P, Villa O, Vishnu A, Vogiatzis KD, Wang D, Weare JH, Williamson MJ, Windus TL, Woliński K, Wong AT, Wu Q, Yang C, Yu Q, Zacharias M, Zhang Z, Zhao Y, Harrison RJ (2020) NWChem: past, present, and future. J Chem Phys 152:184102
67. Lukman S, Musser AJ, Chen K, Athanasopoulos S, Yong CK, Zeng Z, Ye Q, Chi C, Hodgkiss JM, Jishan W, Friend RH, Greenham NC (2015) Tuneable singlet exciton fission and triplet-triplet annihilation in an orthogonal pentacene dimer. Adv Funct Mater 25:5452
68. Kuramochi H, Takeuchi S, Tahara T (2016) Femtosecond time-resolved impulsive stimulated Raman spectroscopy using sub-7-fs pulses: apparatus and applications. Rev Sci Instrum 87:043107
69. Ruhman S, Joly AG, Nelson KA (1988) Coherent molecular vibrational motion observed in the time domain through impulsive stimulated Raman scattering. IEEE J Quant Electron 24:460
70. Zhu L, Sage JT, Champion PM (1994) Observation of coherent reaction dynamics in heme proteins. Science 266:629
71. Liebel M, Schnedermann C, Wende T, Kukura P (2015) Principles and applications of broadband impulsive vibrational spectroscopy. J Phys Chem A 119:9506
72. Loppnow GR, Mathies RA (1988) Excited-state structure and isomerization dynamics of the retinal chromophore in rhodopsin from resonance Raman intensities. Biophys J 54:35

73. Gueye M, Manathunga M, Agathangelou D, Orozco Y, Paolino M, Fusi S, Haacke S, Olivucci M, Leónard J (2018) Engineering the vibrational coherence of vision into a synthetic molecular device. Nat Commun 9:313
74. Stern HL, Musser AJ, Gelinas S, Parkinson P, Herz LM, Bruzek MJ, Anthony J, Friend RH, Walker BJ (2015) Identification of a triplet pair intermediate in singlet exciton fission in solution. Proc Natl Acad Sci 112:7656
75. Liebel M, Schnedermann C, Kukura P (2014) Vibrationally coherent crossing and coupling of electronic states during internal conversion in β-carotene. Phys Rev Lett 112:198302
76. Schnedermann C, Liebel M, Kukura P (2015) Mode-specificity of vibrationally coherent internal conversion in rhodopsin during the primary visual event. J Am Chem Soc 137:2886
77. Haegeman J, Lubich C, Oseledets I, Vandereycken B, Verstraete F (2016) Unifying time evolution and optimization with matrix product states. Phys Rev B 94:165116
78. Evenbly G, Vidal G (2009) Algorithms for entanglement renormalization. Phys Rev B 79:144108
79. Andrzejak M, Skóra T, Petelenz P (2019) Is vibrational coherence a byproduct of singlet exciton fission? J Phys Chem C 123:91
80. Gozem S, Luk HL, Schapiro I, Olivucci M (2017) Theory and simulation of the ultrafast double-bond isomerization of biological chromophores. Chem Rev 117:13502
81. Schapiro I, Ryazantsev MN, Frutos LM, Ferré N, Lindh R, Olivucci M (v) The ultrafast photoisomerizations of rhodopsin and bathorhodopsin are modulated by bond length alternation and HOOP driven electronic effects. J Am Chem Soc 133:3354
82. Del Pino J, Schröder FAYN, Chin AW, Feist J, Garcia-Vidal FJ (2018) Tensor network simulation of polaron-polaritons in organic microcavities. Phys Rev B 98:165416
83. Del Pino J, Schröder FAYN, Chin AW, Feist J, Garcia-Vidal FJ (2018) Tensor network simulation of non-Markovian dynamics in organic polaritons. Phys Rev Lett 121:227401
84. Gélinas S, Rao A, Kumar A, Smith SL, Chin AW, Clark J, van der Poll TS, Bazan GC, Friend RH (2014) Ultrafast long-range charge separation in organic semiconductor photovoltaic diodes. Science 343:512
85. Fuller FD, Pan J, Gelzinis A, Butkus V, Senlik SS, Wilcox DE, Yocum CF, Valkunas L, Abramavicius D, Ogilvie JP (2014) Vibronic coherence in oxygenic photosynthesis. Nat Chem 6:706
86. Liebel M, Schnedermann C, Kukura P (2014) Sub-10-fs pulses tunable from 480 to 980 nm from a NOPA pumped by an Yb:KGW source. Opt Lett 39:4112

Chapter 9
Controlling the Coherent Versus Incoherent Character of Singlet Fission

The results presented in this chapter have been published in Ref. [1] (https://doi.org/10.1021/jacs.9b05561). Adapted with permission from: Antonios M. Alvertis, Steven Lukman, Timothy J. H. Hele, Eric G. Fuemmeler, Jiaqi Feng, Jishan Wu, Neil C. Greenham, Alex W. Chin, Andrew J. Musser. Switching between coherent and incoherent singlet fission via solvent-induced symmetry-breaking. Journal of the American Chemical Society, 141:17558, 2019, Copyright 2019 American Chemical Society.
The experimental work presented in this chapter was not undertaken by the author of this thesis, but by a number of collaborators who are listed as co-authors in the aforementioned publication: Steven Lukman, Jiaqi Feng, Jishan Wu, Neil C. Greenham and Andrew J. Musser. The measurements and their interpretation are an integral part of this combined theoretical-experimental study, and are hence presented here for completeness, along with the associated experimental methodology.

In Chaps. 7 and 8 we employed tensor network methods in order to study the ultrafast exciton dynamics of endothermic charge transfer and exothermic singlet fission in two covalent dimer systems. This allowed us to develop insights on the precise role of molecular vibrations during these processes. We now turn our attention to endothermic singlet fission in the covalent tetracene dimer studied in Chap. 7. The long timescales over which this process occurs (hundreds of picoseconds) prevents us from utilising tensor network methods. Instead, we develop a simple computational model based on physical insights from the previous chapters, including the effects of molecular vibrations, excess energy and the dielectric environment of a molecule. This not only allows us to interpret complex experimental data on the singlet fission dynamics of the studied system, but also to show for the first time that it is possible to enter the so-called coherent regime of quantum dynamics in a controlled way.

© The Author(s), under exclusive license to Springer Nature Switzerland AG 2021
A. M. Alvertis, *On Exciton–Vibration and Exciton–Photon Interactions in Organic Semiconductors*, Springer Theses,
https://doi.org/10.1007/978-3-030-85454-6_9

9.1 Introduction

Singlet fission [2] is an electronic process in organic materials which has been extensively studied in the past decade. This is largely because of its promise for efficient solar energy technologies which surpass the Shockley–Queisser limit [3]. Singlet fission converts a high-energy singlet exciton to two low-energy triplet excitons, which are at least initially coupled into an overall singlet state. Thus singlet fission conserves spin, allowing it to be an ultrafast process and effectively compete with radiative and non-radiative deactivation of excited states. This energetic down-conversion process offers a way to overcome thermalisation losses, and has inspired the design of new hybrid device concepts [4, 5]. However, the primary interest in the field to date remains building a more detailed understanding of the underlying photophysical mechanism, with the aim of informing rational materials design.

There have been multiple reports that singlet fission can occur through an ultrafast (fs-ps) 'coherent' mechanism [6–10]. Here, photoexcitation is thought to generate an initial superposition of the lowest bright singlet state S_1 and the dark double-triplet state TT which eventually dephases into the dark state. In some materials, this coherent regime is reported to coexist with slower, incoherent fission dynamics [9, 10], in which S_1 'hops' to the TT surface non-adiabatically, or adiabatically relaxes into the TT region on the same potential energy surface. Incoherent singlet fission is the more commonly invoked picture, particularly when it occurs on longer ($>$ ps) timescales [11, 12]. Interestingly, even in systems where it is endothermic, e.g. in tetracene [13–16], singlet fission can be very efficient and temperature-independent [15–18]. To explain such observations, coherent processes have been invoked [10, 19].

Despite recent progress in understanding the coherent mechanism of singlet fission and its interplay with incoherent dynamics, what has thus far been missing is a tuneable way of switching between the two regimes. This is partly because there are no established experimental handles to achieve this, and the utility of coherent dynamics as a concept for materials design and eventual applications is not clear. Similarly to many other photophysical processes in organic molecules, one of the most promising angles to explore coherent dynamics is through molecular vibrations. These are increasingly recognised both in experiment [10, 17, 20, 21] and theory [22–27] to play a critical role in the ultrafast regime where singlet fission is often found to occur. The coherent mechanism of singlet fission in rubrene (which has a well-defined symmetry) has been shown to arise via symmetry-breaking modes that allow S_1 and TT to mix [10]. More broadly, when the mixing between electronic states in such systems is governed by coupling to intramolecular motions, to some extent it can be tuned, for example by controlling the viscosity of the environment [28] or chemically introducing steric barriers [29]. This control is rather limited and poorly understood, especially in solid-state systems where the complex interplay of intra- and intermolecular vibrations must be considered.

Another critical ingredient in both coherent and incoherent fission mechanisms that has proved most difficult to explore is the nature of the coupling that results in quantum superpositions or population transfer between S_1 and TT. There has been extensive debate about the relative strength and influence of 'direct' two-electron coupling between these states, versus sequential one-electron couplings mediated by a charge-transfer (CT) state [2, 12, 30–33]. Couplings involving the CT state are typically expected to be orders of magnitude higher, but in conventional thin-film systems there is no experimental means to perturb the CT manifold, especially if the states are not directly populated but only mediate fission through super-exchange as 'virtual' intermediates [30, 33]. However, covalent dimers of singlet-fission chromophores offer exquisite tunability of the critical interactions that govern fission, both through chemical design [34–44] and the external environment [28, 45, 46], and they thus present the most promising platform to describe and control such effects. In these systems, the energies of CT states can be directly tuned through the use of different solvents. This concept has enabled the demonstration of fission mediated by direct S_1-TT coupling [29], virtual CT states [27, 28, 45, 46] and even a distinct CT intermediate [28]. In short, covalent dimers offer a versatile platform to systematically explore the S_1-TT transition and understand the nature of the coherent/incoherent pathways.

Here, we combine a number of theoretical techniques with an experimental investigation based on ultrafast spectroscopy in order to reveal the detailed mechanism of singlet fission in the orthogonal tetracene dimer DT-Mes (Fig. 9.1a) in solution and demonstrate how it is governed by a tuneable 'switch' between the incoherent and coherent regimes. The experimental findings show that singlet fission in this system is a 'hot' process, i.e. it only occurs upon excitation distinctly above the band-edge of the first excited singlet state [47]. Its mechanism is found to be qualitatively different depending on the solvent environment, and the fission efficiency is maximised in intermediate polarity solvents. Our theoretical analysis allows us to identify the crucial role of molecular vibrations to the singlet fission process, though alone they only give rise to weak mixing between states participating in singlet fission, and can only account for the experimental observations in low-polarity solvents where the process is incoherent. In more polar solvents we observe a qualitatively different regime of singlet fission, and we can only capture this behaviour through the inclusion of dynamic solvent effects. Our model reveals that increasing the polarity switches DT-Mes into a regime of coherent singlet fission. However, contrary to previous studies reporting on superpositions of S_1 and TT that dephase into the triplet pair [7, 8, 10, 19], here we show that the coherent mechanism comes into play through the mixing of a CT state and TT. Coherent singlet fission occurs in solvents of intermediate polarity, where dynamic solvent effects induce the closest energetic proximity between these states so that molecular vibrations can mix them most efficiently. It is precisely the properties of CT states which allow us to tune them through the dielectric environment, which in contrast to local-excitations and triplet pairs, have a finite electric dipole. This underlines the novelty of our 'switch' mechanism, and emphasises the potential of using CT states to tune coherent dynamics.

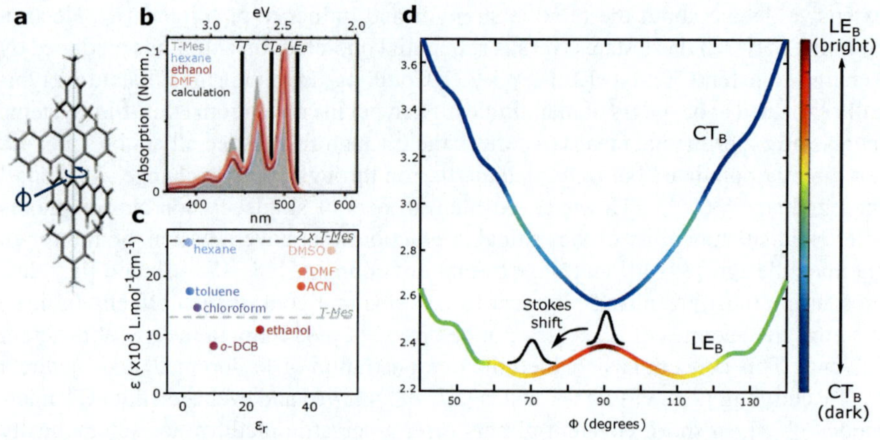

Fig. 9.1 **a** Chemical structure of DT-Mes in its orthogonal ground-state geometry, indicating the dominant torsional motion. **b** Ground state absorption spectra of DT-Mes in solvents of different polarity and mesityl-substituted tetracene monomer T-Mes in chloroform (shaded). Calculated electronic energy levels are shown as vertical bars. The energy of the bright state LE_B is in excellent agreement with the position of the absorption onset. **c** Peak molar extinction coefficient of DT-Mes in different solvents compared to that of T-Mes in chloroform (dashed). The extinction coefficient is minimised in intermediate polarity solvents. **d** Potential energy surfaces of the two bright excited states along the torsional coordinate. The colour code indicates the mixing between the LE_B and CT_B surfaces, leading to finite oscillator strength for the latter through intensity borrowing. The maximal mixing is found at the local minima of LE_B, from where strongly red-shifted emission occurs

Results and Discussion

9.2 Molecular Structure and Coupling

DT-Mes (chemical structure in Fig. 9.1a) consists of two tetracenes directly linked at the 5,5′ position, with mesityl side groups for solubility. Similarly to the anthracene [48] and pentacene [28] analogues and a closely related cyano-substituted tetracene dimer [49], DFT geometry optimisation reveals that steric repulsion forces the two tetracenes into an orthogonal geometry. Being an alternate hydrocarbon, DT-Mes has no permanent dipole in the ground electronic state, as predicted from the Coulson–Rushbrooke theorem [50]. Therefore, the ground state geometry does not depend on the dielectric environment.

The orthogonal geometry raises the question of whether the two tetracenes can interact electronically, given their negligible π overlap [28, 45]. The presence and relative strength of interchromophore coupling is determined from the steady-state absorption (Fig. 9.1b). This does not directly map onto the excited-state couplings between dark states (CT and TT) actually relevant to singlet fission [39], but it serves as a useful proxy [11]. Compared to the equivalent mesityl-tetracene monomer (shaded), the dimer (lines) exhibits a more prominent 0–0 absorption peak, though the exact 0–0/0–1 ratio varies with solvent. The enhanced 0–0 peak reflects J-type excitonic coupling between the short-axis-polarised $S_0 \rightarrow S_1$ transitions of the tetracenes. Moreover, the solvent dielectric constant has a powerful effect on the molar extinction coefficient (Fig. 9.1c). In the extremes of the solvent polarity series we find that the dimer absorbs two times as strongly as the monomer. This would be consistent with negligible inter-tetracene interactions, similar to several prior reports of weakly coupled dimers [35–38]. However, in intermediate solvents the extinction coefficient of the dimer is less than that of even a single monomer, reaching as low as 40% of the expected value. This strong hypochromism is compelling evidence for strong interchromophore coupling, presumably mediated by interaction with dark charge-transfer (CT) states [28] given the strong solvent dependence. This is further supported by the model approach which is developed in this chapter (see Sect. 9.7), which reproduces the behaviour of the molar extinction coefficient in different solvents due to mixing with the CT states.

As in the equivalent pentacene dimers [28, 45], we propose that the tetracenes are simultaneously coupled through CT and excitonic interactions. These have opposite effects on the energy of the bright state (a red-shift from the excitonic coupling and a blue-shift from the CT-mediated coupling), and the combination results in 'null aggregates' with weak spectral shifts and small changes in vibronic structure despite the significant interactions [51]. Within this framework, the results in Fig. 9.1c demonstrate that CT states play a central and tuneable role in the dimer electronic structure.

9.3 Exciton States and State Mixing

The ground state geometry of DT-Mes was optimised using DFT with the B3LYP functional and cc-pVDZ basis set, and it was found to assume a 90° configuration of the central dihedral angle. There are two competing effects, the interplay of which determines the molecular structure: orbital delocalisation and steric repulsion between the neighbouring hydrogen atoms of the two tetracene monomers. For such a large structure, strong steric repulsions dictate the orthogonal configuration, see Fig. 9.1. The structure is C_2 symmetric, therefore the molecular orbitals will transform either as the A or B irreducible representation of the point group. By labelling the two tetracene monomers as m and n, the dimer frontier molecular orbitals may be written as linear combinations of the monomer ones:

$$
\begin{aligned}
|H-1\rangle &= \frac{1}{\sqrt{2}}(|h_m\rangle + |h_n\rangle) \quad (A \text{ symmetry}) \\
|H\rangle &= \frac{1}{\sqrt{2}}(|h_m\rangle - |h_n\rangle) \quad (B \text{ symmetry}) \\
|L\rangle &= \frac{1}{\sqrt{2}}(|l_m\rangle + |l_n\rangle) \quad (A \text{ symmetry}) \\
|L+1\rangle &= \frac{1}{\sqrt{2}}(|l_m\rangle - |l_n\rangle) \quad (B \text{ symmetry}).
\end{aligned} \tag{9.1}
$$

In order to approximate the energy of the singlet TT state ^{1}(TT) for DT-Mes, we calculated the energy of the ground state quintet ^{5}(TT), similar to Chap. 8, and obtained $E(\text{TT}) = 2.78$ eV.

For calculating the excited singlet states of the system we use Pople–Parr–Pariser (PPP) theory [52, 53]. These calculations are known to be very successful in predicting the absorption spectra of π systems and include correlation effects. In describing the electronic structure and properties of DT-Mes, we use as our basis the five lowest-energy adiabatic excited states in the orthogonal ground-state geometry, calculated in the absence of solvent effects. These coincide with the diabatic states relevant for singlet fission, and the first four are the symmetric and antisymmetric linear combinations of the excitations localised on the two monomers $\text{LE}_{A,B}$, and the symmetric and antisymmetric linear combinations of the two CT states, $\text{CT}_{A,B}$:

$$
\begin{aligned}
|\text{LE}_A\rangle &= \frac{1}{\sqrt{2}}(|\text{LE}_m\rangle + |\text{LE}_n\rangle) \quad (A \text{ symmetry}) \\
|\text{LE}_B\rangle &= \frac{1}{\sqrt{2}}(|\text{LE}_m\rangle - |\text{LE}_n\rangle) \quad (B \text{ symmetry}) \\
|\text{CT}_A\rangle &= \frac{1}{\sqrt{2}}(|\text{CT}_{m\to n}\rangle + |\text{CT}_{n\to m}\rangle) \quad (A \text{ symmetry}) \\
|\text{CT}_B\rangle &= \frac{1}{\sqrt{2}}(|\text{CT}_{m\to n}\rangle - |\text{CT}_{m\to n}\rangle) \quad (B \text{ symmetry}).
\end{aligned} \tag{9.2}
$$

The transitions always refer to an electron transfer from the HOMO of a monomer to the LUMO of the same or the second monomer. These diabatic states are symmetry-adapted linear combinations, spanning the A and B irreducible representations of the C_2 point group. The effect of a C_2 rotation on a monomer-localised excitation is visualised in Fig. 9.2. Since in acenes such localised states are short-axis (i.e. y-axis here) polarised, the C_2 rotation results in a change of sign of the transition dipole moment. Therefore, from the constructed symmetry-adapted linear combinations of localised states, $|\text{LE}_B\rangle$ will have a non-vanishing transition dipole moment in the y direction, while $|\text{LE}_A\rangle$ will be dark, as the transition dipoles of the two localised states cancel out, see Fig. 9.3. Both the $|\text{CT}_{m\to n}\rangle$ and $|\text{CT}_{n\to m}\rangle$ states are dark, therefore the defined diabatic CT states are also dark. Hence it is only $|\text{LE}_B\rangle$ which is bright from the above.

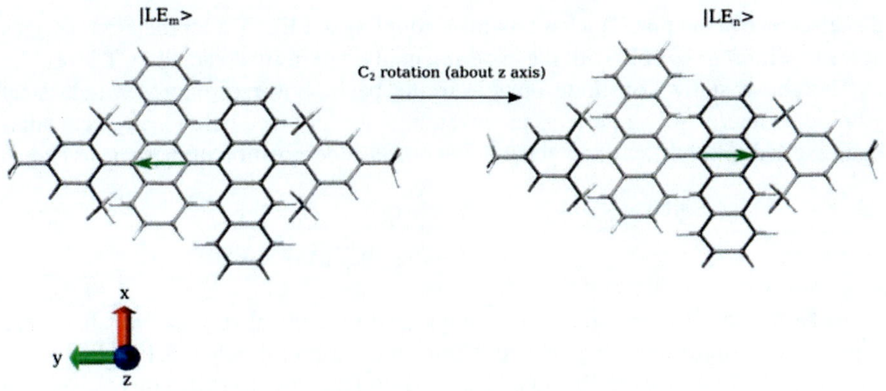

Fig. 9.2 Application of a C_2 rotation on a monomer-localised short-axis (y) polarised excitation of DT-Mes results in a localised excitation of opposite transition dipole moment

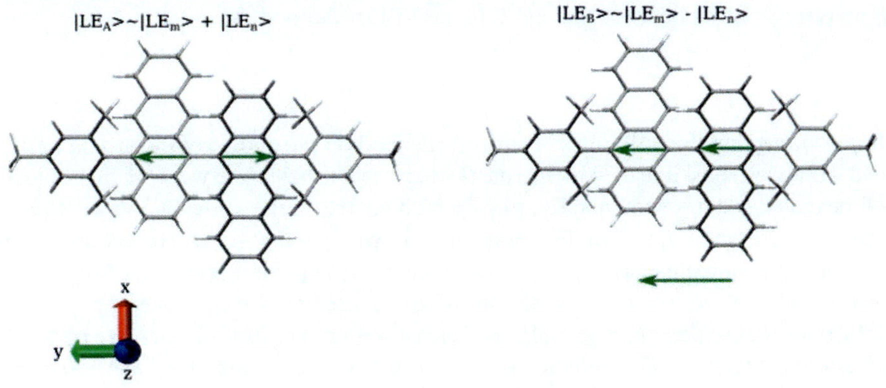

Fig. 9.3 Constructing the transition dipoles of $|\mathrm{LE}_A\rangle$ and $|\mathrm{LE}_B\rangle$; the earlier vanishes, while the latter is finite

Having obtained the ground state structure (i.e. 90° of angle), and subsequently the first four adiabatic states and TT, one can write the electronic Hamiltonian in the diabatic basis $\{\mathrm{LE}_B, \mathrm{LE}_A, \mathrm{CT}_B, \mathrm{CT}_A, \mathrm{TT}\}$ in eV:

$$H_{el} = \begin{bmatrix} 2.39 & 0 & -0.014 & 0 & 0 \\ 0 & 2.56 & 0 & 0 & 0 \\ -0.014 & 0 & 2.57 & 0 & 0 \\ 0 & 0 & 0 & 2.57 & 0 \\ 0 & 0 & 0 & 0 & 2.78 \end{bmatrix}, \tag{9.3}$$

from where it becomes apparent that singlet fission is endothermic in this system. The calculated LE_B energy of 2.39 eV (vertical bar in Fig. 9.1b) is in very good agreement with the experimental value of 2.48 eV. The two CT states are degenerate, the difference being that CT_B has a 14 meV coupling to LE_B. Their predicted energies agree to within ≈ 0.1 eV with the experimentally determined vacuum CT level.

The above states constitute our electronic basis, however once semi-classical vibrational and/or solvent effects are taken into account (see below), a typical adiabatic excited state $|\Psi\rangle$, will, in general, become a superposition of these basis states:

$$|\Psi\rangle = c_{\mathrm{LE}_B}\,|\mathrm{LE}_B\rangle + c_{\mathrm{LE}_A}\,|\mathrm{LE}_A\rangle + c_{\mathrm{CT}_A}\,|\mathrm{CT}_A\rangle + \\ + c_{\mathrm{CT}_B}\,|\mathrm{CT}_B\rangle + c_{\mathrm{TT}}\,|\mathrm{TT}\rangle \tag{9.4}$$

To make it easier to compare spectroscopic and theoretical results, we label these superposition states according to their dominant contribution, e.g. if $|\mathrm{LE}_B\rangle$ dominates the sum of Eq. (9.4), we call the state LE_B for simplicity. We stress that this convention is not meant to imply that the state $|\Psi\rangle$ is exactly equal to one of the basis states, which are only present in their pure form in the ground-state geometry at low polarity. For such a superposition state $|\Psi\rangle$, the mixing (or coherence as defined in Chap. 3) between two of its basis states, e.g. CT_B and TT, is defined as:

$$\rho_{\mathrm{CT}_B\mathrm{TT}} = c^*_{\mathrm{CT}_B} \cdot c_{\mathrm{TT}}. \tag{9.5}$$

It is obvious that this mixing will strongly depend on the choice of electronic basis, and it is always possible to work in a basis where it vanishes. Here we work in the basis of the diabatic states shown in Eq. (9.4), which are themselves equal to the *adiabatic* states at the Franck–Condon (FC) point and in a non-polar solvent. Therefore, initial photoexcitation forms a pure LE_B state, and any subsequent mixing occurs between states that were originally separate, and is not an artefact of our choice of basis.

To investigate the effect of molecular vibrations on the photophysics, we perform electronic structure calculations for a range of displacements along the torsional angle Φ between the monomers (Fig. 9.1a). One needs to take care at this point: PPP calculations do not take steric effects into account, i.e. they give the energy gaps between ground and excited states for a given geometry, but not the energy difference between different geometries. We therefore used ground-state DFT to obtain relaxed DT-Mes geometries, with PPP calculations performed 'on top' of that to account for mesomeric effects. The motion along Φ crucially underpins the photophysics of both the anthracene [48] and pentacene [28] analogues. In DT-Mes, this is a low-frequency motion (39.6 cm^{-1}) of A symmetry. From analysing the excitation character of the surfaces, we conclude that displacement along Φ leads to mixing between bright LE_B and dark CT_B (Fig. 9.1d). This results in CT_B borrowing intensity from LE_B and becoming partially bright [54]. Our calculations also reveal energetic relaxation of LE_B with increasing Φ and away from the FC point, to a minimum at 70°. This tendency towards planarisation gives rise to Stokes-shifted emission (see below) and is also documented in the equivalent anthracene [48] and pentacene [28] dimers.

In a final clarification about notation, we point out that solvent effects result in mixing between the two symmetry-pure CT states. This results in the formation of two new CT states which are linear combinations of the originals. One of these has lower energy, which we denote as $CT_{\downarrow}$, while the other one is destabilised and denoted as $CT_{\uparrow}$. Both of these have finite CT_B character, hence both borrow intensity from LE_B and are partially bright, with the potential to emit photons. However, in practice we only detect photon emission from $CT_{\downarrow}$. For ease we use this as the primary label for CT-related experimental signatures, with the recognition that it denotes a solvent-dependent mixture of CT_B and CT_A. We discuss the nature and implications of this mixing in greater detail below, following the presentation of experimental data.

9.4 Multiple Emissive Species

The energetic landscape of DT-Mes may be defined through photoluminescence spectroscopy. Though the steady-state absorption spectra of monomer and dimer are almost identical (Fig. 9.1a), the steady-state photoluminescence of DT-Mes in solution (Fig. 9.4a, solid lines) exhibits a large Stokes shift of $\approx$ 35 nm not observed in the monomer (shaded spectrum, $\approx$ 2 nm. This shift can be reduced in high-viscosity polystyrene solution (dashed spectrum), confirming that it is linked to large-scale conformational change such as relaxation along Φ. These measurements also show a more pronounced solvent dependence than the steady-state absorption. In the four least-polar solvents the emission is dominated by a well-defined vibronic progression 500–600 nm which does not shift and can be assigned to the relaxed (i.e. partially planarised) singlet $LE_{B\mathrm{rel}}$. In the more polar solvents, the emission becomes featureless and progressively red-shifts with increasing polarity. This behaviour is a hallmark of CT state emission [28, 55]. Similar features are also detected at the high-energy edge in non-polar solvents (inset), i.e. at shorter wavelengths than LE_B photoluminescence. This non-Kasha emission, from a state which is not the lowest-energy singlet in the molecule, is surprising, though long-lived high-energy CT states are known in similar orthogonal systems [56]. Here, it indicates the presence of multiple emissive species (LE_B and $CT_{\downarrow}$) in DT-Mes in non-polar solvents.

In non-polar solvents, it is possible to tune the balance of these species through the pump photon energy. Band-edge excitation in hexane yields purely excitonic emission, with a quantum efficiency of $\approx$ 68% (Fig. 9.4b). With increasing photon energy, we detect a greater proportion of non-Kasha $CT_{\downarrow}$ emission. This changing balance is accompanied by a corresponding decrease in the photoluminescence quantum yield until saturation at $\approx$ 20%. This is comparable to the quantum yield when the emission is dominated by $CT_{\downarrow}$, as observed in all polar solvents following band-edge excitation. Interestingly, the behaviour of the emission quantum efficiency qualitatively changes for excitation above the pump photon energy threshold of 2.8 eV both for polar and non-polar solvents—monotonic decrease and saturation, respectively. This behaviour suggests a significant change in photophysical processes above this energy.

To guide the analysis of these excitation-dependent processes, the essential results from photoluminescence measurements are first compiled, in order to describe the DT-Mes energetic structure. Time-correlated single-photon counting reveals that in

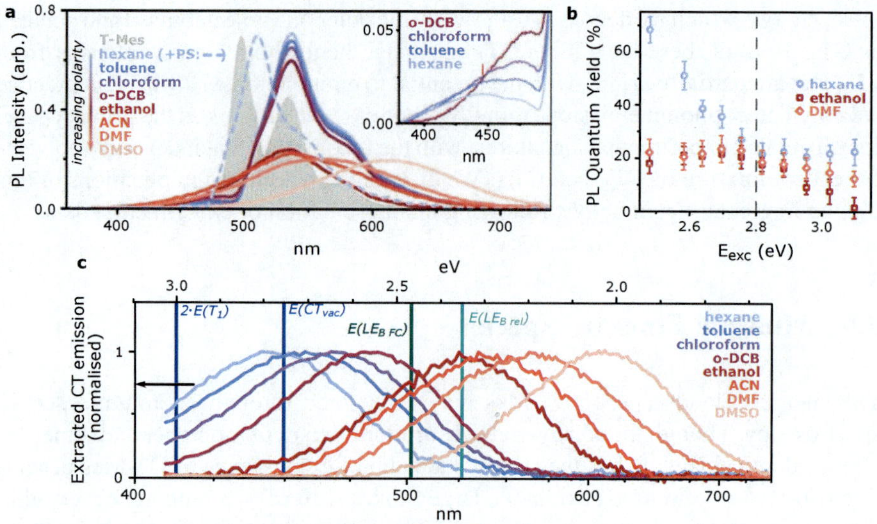

Fig. 9.4 **a** Steady-state photoluminescence of DT-Mes in a range of solvents (solid lines), following excitation at 3.1 eV. The red-shift relative to monomer T-Mes (grey shaded) can be suppressed in high-viscosity solution (dashed). Inset: non-Kasha emission observed in non-polar solvents. **b** Solvent- and excitation-dependent photoluminescence quantum yield. In all media, the behaviour of the system changes markedly for excitation at energies above 2.8 eV. **c** Energy landscape constructed from emission measurements. $CT_\downarrow$ spectra (lines) are extracted from time-resolved emission measurements in each solvent. Vertical bars denote the energies of other relevant electronic states. LE_{BFC} and LE_{Brel} are determined in non-polar solvents with and without polystyrene. $E(T_1)$ is determined from sensitised phosphorescence and $E(CT_{vac})$ is extrapolated from the CT solvent dependence (see Fig. 9.7). The energies of all emissive states are below the presumed singlet fission threshold $2 \cdot E(T_1)$

non-polar solvents the non-Kasha $CT_\downarrow$ emission has a lifetime of $\approx$ 20 ns. This is comparable to observations in polar solvents, where $CT_\downarrow$ is the only emissive species. This is significantly longer than the excitonic LE_B emission observed in polar solvents, which has a lifetime of $\approx$ 6 ns. This large difference allows straightforward spectral decomposition to isolate the $CT_\downarrow$ emission spectra, plotted for all solvents in Fig. 9.4c. This emission is compared to the energies of key electronic states (vertical bars). Unsurprisingly, examination of the $CT_\downarrow$ and excitonic energies reveals that CT emission dominates in the four polar solvents in which $CT_\downarrow$ is the lowest-energy state. Comparing to Fig. 9.1c, we also find that the minimum in extinction coefficient coincides with the point at which $E(CT_\downarrow)$ is closest to $E(LE_{BFC})$. This energetic proximity presumably enables the strongest CT-mediated coupling between tetracenes. Interestingly, sensitised phosphorescence measurements show that the dimerisation motif of DT-Mes substantially destabilises the triplet state from 1.3 to 1.5 eV, which is denoted by the vertical bar at 3 eV. This is a surprising effect which was also observed in orthogonal pentacene dimers [28], and its origin is not currently understood. We recall, though, that previous studies on perylene diimide films found the triplet energy varies with intermolecular coupling [57], and significant energy shifts have been reported between TIPS-tetracene [58] and a phenyl-substituted derivative [59]. There is evidently significant scope—currently little explored—to tune the triplet energy on the same parent chromophore. The crucial result of these measurements for our purposes here is that none of the states observed in absorption or emission approach the energy expected to be required for singlet fission. Nor is it immediately evident what is the origin of the observed thresholding behaviour appearing at 2.8 eV in Fig. 9.4b, since it would not be expected to arise from singlet fission.

9.5 CT-Mediated Singlet Fission

To better understand the system's behaviour and the role of dark electronic states, the dynamics of DT-Mes have been probed using transient absorption spectroscopy in all eight solvents with ten pump photon energies, spanning from band-edge excitation 2.48 eV to significant excess energy 3.10 eV. We present representative results for ethanol solution in Fig. 9.5. Below the 2.8 eV excitation threshold (Fig. 9.5a) one can identify the signatures of three distinct electronic states. The initial state $\mathrm{LE}_{B\mathrm{FC}}$ is characterised by a prominent trio of stimulated emission ($\Delta T/T > 0$) peaks at 510, 540 and 580 nm and a well-defined excited-state absorption ($\Delta T/T < 0$) at 630 nm. These signatures evolve with $\approx$ 550 fs time-constant to a state with strongly attenuated stimulated emission and a broader, flattened excited-state absorption. In corresponding transient grating photoluminescence [60] measurements, a pronounced redshift of the emission is observed on precisely the same timescale, consistent with geometric relaxation of the excited state. This 550 fs time constant is intermediate between the planarisation dynamics reported for the equivalent anthracene [61] and pentacene [28] dimers. Therefore, the second species in transient absorption can be assigned to the partially planarised singlet $\mathrm{LE}_{B\mathrm{rel}}$. On longer timescales a new species is evident with unique excited-state absorption in the near-infrared (e.g. 710, 775, 880 nm). This state is similarly long-lived to the $\mathrm{CT}_{\downarrow}$ emission ($\approx$ 20 ns) and the features in the near-infrared closely match peaks observed in the chemical oxidation and reduction spectra (Fig. 9.5c), allowing assignment to the CT state. While it contains multiple radical anion and cation signature peaks, the imperfect match in Fig. 9.5c also shows that CT is not a pure $\mathrm{D}^+\mathrm{A}^-$ state. The deviations from the $\mathrm{D}^+\mathrm{A}^-$ spectrum may be attributed to LE contributions to the total wavefunction, which also give the state the ability to emit. In short, a simple excited-state progression is observed, from the initial bright singlet state $\mathrm{LE}_{B\mathrm{FC}}$ to a conformationally relaxed singlet $\mathrm{LE}_{B\mathrm{rel}}$, and from there to a long-lived emissive CT state. Similar dynamics are observed in all solvents, for excitation below 2.8 eV but above the energy of the $\mathrm{CT}_{\downarrow}$ state (which varies by solvent). No signatures of triplet excitons (Fig. 9.5c, bottom) are detected in these conditions, demonstrating that singlet fission is inactive and intersystem crossing from LE and CT states is inefficient.

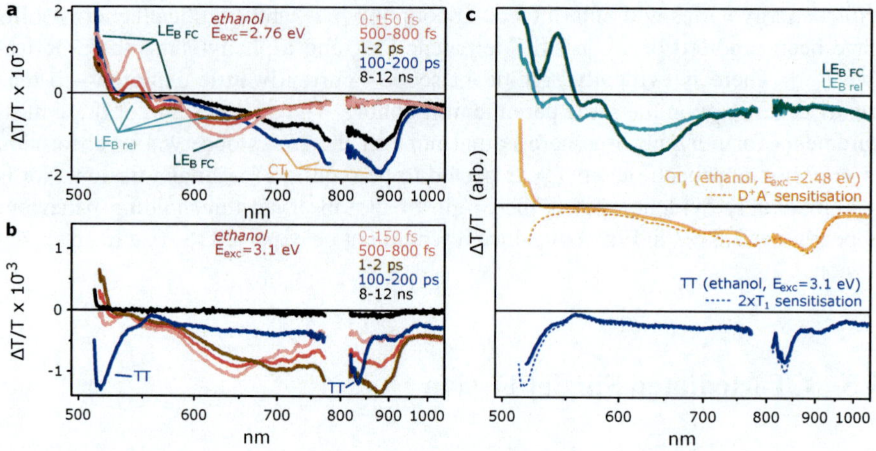

Fig. 9.5 **a** Transient absorption spectra of DT-Mes in ethanol for excitation below the 2.8 eV threshold. Signatures of LE_{BFC}, LE_{Brel} and $CT_{\downarrow}$ are indicated. **b** $E_{exc} > 2.8$ eV generates a different initial state from lower-energy excitation. This state evolves into the same $CT_{\downarrow}$ signatures 1–2 ps observed in **a**. Subsequent $CT_{\downarrow}$ decay yields distinctive TT signatures (indicated), which largely decay < 1 ns. **c** Species-associated spectra extracted from transient absorption spectroscopy (solid) and reference spectra from charge and triplet sensitisation (dashed). The relative magnitudes of the spectra reflect their relative molar extinction coefficients

For excitation above the 2.8 eV threshold, Fig. 9.5b, the initial excited-state absorption signature is distorted from that of LE_{BFC}, exhibiting much weaker stimulated emission. Nonetheless, it rapidly evolves to the same CT state identified in Fig. 9.5a. The latter now undergoes a new decay pathway. At 100–200 ps the unique excited-state absorption fingerprint of DT-Mes triplets at 520 and 830 nm is detected, identified from solution sensitisation (Fig. 9.5c). The lifetime of these triplets is ≈ 800 ps, 5000 times shorter than the lifetime of individual triplets in sensitisation. We are only able to explain such a significant reduction in the triplet lifetime through triplet-triplet annihilation. At the low concentrations and excitation densities used in the experiments, that requires two triplet excitons to be produced on a single DT-Mes molecule, ruling out intersystem crossing from LE states [62] or triplet generation from CT recombination [63] as possible formation mechanisms. Instead, this rapid annihilation of triplet pairs is considered to be a hallmark of intramolecular singlet fission, as previously observed in numerous dimer and conjugated-polymer studies [28, 35, 46, 64–66]. In the case of DT-Mes, the fission process is mediated by a CT state which is directly populated and presents distinct spectroscopic features. Though often predicted [2, 32], this fission mechanism has proven experimentally elusive and has only been conclusively identified in one other system, a similar orthogonal pentacene dimer with TIPS solubilising groups [28]. It is far more common for CT states to modulate singlet fission through a 'virtual' or superexchange pathway [27, 28, 30, 45, 46], but the directly observable CT intermediate here allows us to obtain deeper mechanistic insight.

9.6 Singlet Fission Yield Variation

The basis spectra identified in Fig. 9.5c are found to be sufficient to describe the excited-state progression in all solvents and at all pump photon energies, enabling easy comparison between experimental conditions. Figure 9.6 highlights how the terminal state branches between fission-generated TT and long-lived CT at 100–200 ps. At this time delay singlet fission—when it occurs—is complete. In Fig. 9.6a one observes that the TT yield is clearly optimised for intermediate solvents such as ethanol and *o*-DCB. These are the solvents where the CT and LE_B energies most closely approach (Fig. 9.4c) and where the strongest LE-CT coupling is inferred from the low oscillator strength (Fig. 9.1c). For solvents with higher or lower polarity, increased branching into the long-lived $CT_\downarrow$ state is detected. Moreover, in ethanol the branching into TT increases monotonically with pump photon energy up to a maximum of 3.1 eV (panel b, instrument limit). The TT yield is thus a complex function of both excitation energy and solvent. The optically activated singlet fission reported in Fig. 9.6b is particularly unusual. It has never before been observed in covalent dimers, and while it recalls the behaviour of certain conjugated polymers [47, 66, 67], the important distinction is that the activated process is not fast. Indeed, in some of the solvents singlet fission proceeds over tens of ps, a remarkably long timescale for 'hot' dynamics, which we discuss further below.

Using the excited-state signatures identified in Fig. 9.5c, the yields of TT and long-lived CT obtained in all of the transient absorption datasets can be determined. These are presented in the action spectra in Fig. 9.7a, which summarise the full

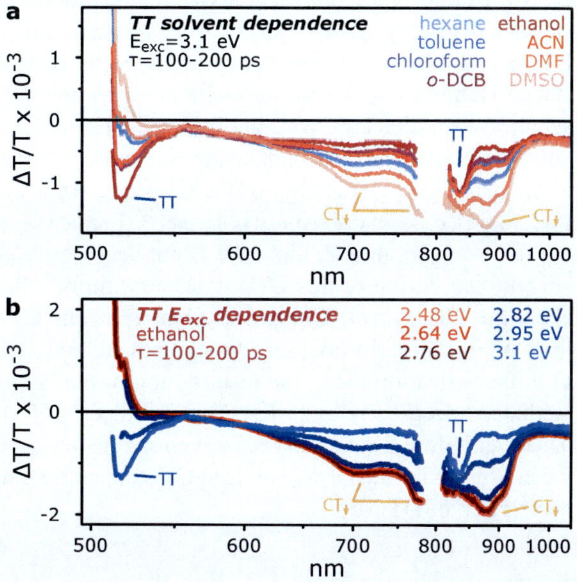

Fig. 9.6 **a** Solvent dependence of TT yield at 100–200 ps, following excitation at 3.1 eV. The TT yield changes depending on the dielectric environment, and is maximised in intermediate polarity solvents. **b** Pump photon energy dependence of TT yield at 100–200 ps in ethanol. TT forms for excitation above 2.8 eV, and its yield increases with excitation energy up to 3.1 eV. The primary features of TT and $CT_\downarrow$ are indicated again for clarity

solvent- and excitation-dependent behaviours highlighted in Fig. 9.6. The orange-shaded spectra reproduce the CT emission from Fig. 9.4c. In polar solvents where $E(\mathrm{CT}_\downarrow) < E(\mathrm{LE}_{BFC})$ (right), the CT action spectra (circles) reveal that initial conversion into CT is always quantitative. In the non-polar solvents (left), the threshold for CT formation roughly follows the envelope of $\mathrm{CT}_\downarrow$ emission, suggesting that this emission lineshape is a reasonable proxy for the electronic energy of the state. This result demonstrates that if the initial excitation has greater energy than some portion of the broader $\mathrm{CT}_\downarrow$ distribution, then the system will access that state. We note that in non-polar solvents the total yields do not sum to 100%. In these solvents the initial formation of $\mathrm{CT}_\downarrow$ competes with relaxation into $\mathrm{LE}_{B\mathrm{rel}}$, which can be detected in transient absorption and the vibronically structured emission (Fig. 9.4a). This state exhibits a lifetime of $\approx$ 6 ns and evidently does not undergo singlet fission. The yield of $\mathrm{LE}_{B\mathrm{rel}}$ is equivalent to the difference between the presented yields and 100%.

In every solvent, regardless of the threshold energy for CT formation, we observe the same threshold energy for singlet fission of ≈ 2.8 eV. It is noteworthy that this energy is significantly lower than the expected fission threshold of $2 \cdot E(\mathrm{T}_1) \approx 3.0$ eV. It is common to invoke an entropic driving force to explain such endothermic fission in solid-state systems [17, 19, 68], but that should not be a factor in a strictly dimeric system. Instead, we can only rationalise this low onset energy through a binding energy. The results imply that the immediate product of singlet fission is a bound triplet-pair state [17, 69–71]. The spin-singlet triplet pair can be significantly stabilised relative to two 'free' triplets due to mixing of the diabatic TT wavefunction with other LE and CT configurations [25, 71, 72]. While the same bound TT state is formed in all cases, the nature of the threshold appears to change from non-polar (Fig. 9.7, left) to polar (Fig. 9.7, right) solvents. In non-polar solvents the TT yield is nearly constant above the threshold. In polar solvents, the TT yield shows a distinctly gradual onset. We can explain this phenomenon using the pair of CT states illustrated schematically in Fig. 9.7b. Here we recall that the emissive CT state is stabilised by interactions with the solvation shell (bottom), hence it red-shifts as the solvent dielectric constant increases. From the solvent dependence of the CT emission we can then extrapolate the vacuum CT energy as 2.75 eV: this corresponds to the bare electronic energy of the CT state. We propose that for a given solvation shell there exists an oppositely polarised CT state (top) which will be destabilised by the same degree, though that state is not necessarily populated. To approximate the energy distribution of this $\mathrm{CT}_\uparrow$ state, we simply reflect the CT emission spectrum about the $\mathrm{CT}_{\mathrm{vac}}$ level. This process generates the $\mathrm{CT}_\uparrow$ spectrum we plot in dashed lines in Fig. 9.7a. Surprisingly, this graphical approach provides a remarkably good fit to the activation data. The results suggest that singlet fission in polar solvents is mediated specifically by a destabilised 'upper' CT state, which to our knowledge has never before been observed or even suggested. To understand this behaviour and the possible role such a destabilised CT state could play, we return to our theoretical description of DT-Mes.

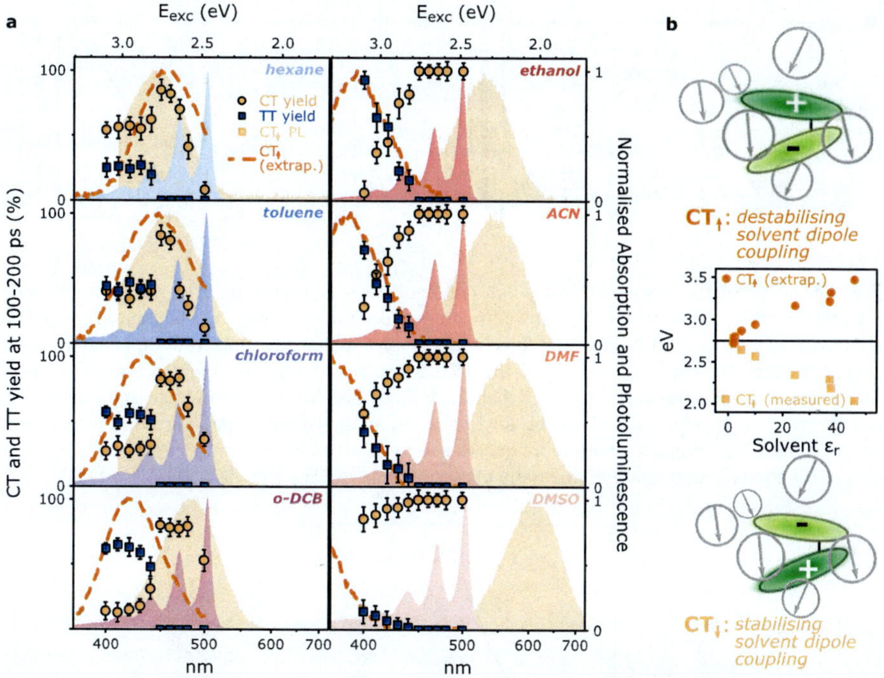

Fig. 9.7 **a** Yield of long-lived $CT_{\downarrow}$ (circles) and TT pairs (squares), extracted from excitation-dependent transient absorption measurements at 100–200 ps time delay, for solvents of increasing polarity (left to right). The TT cross-section was determined by triplet sensitisation, assuming TT $\approx 2 \cdot T_1$. The $CT_{\downarrow}$ yield is relative to the yield in ethanol following band-edge excitation, which we assume to be 100%. Shaded spectra are steady-state absorption and extracted $CT_{\downarrow}$ emission from Fig. 9.4c. Dashed spectra approximate the $CT_{\uparrow}$ distribution, by reflecting the $CT_{\downarrow}$ emission spectrum across the vacuum CT energy CT_{vac} (see main text and panel **b**). **b** Peak of CT emission (squares) as a function of solvent ϵ_r allows extrapolation of the vacuum CT energy ≈ 2.75 eV (blue). $CT_{\downarrow}$ is stabilised from this level by interactions with the solvent shell (bottom), and the oppositely polarised state would be equally destabilised (top), giving rise to the $CT_{\uparrow}$ state (circles)

9.7 Modelling of Vibrational and Solvent Effects

The experimental results of the previous sections show that vibrations must be considered to explain the excited state photophysics of DT-Mes: the first step observed in transient absorption measurements is relaxation along Φ (Fig. 9.5), and suppressing this channel through the use of polystyrene has a significant effect on the emission properties (Fig. 9.4a). Furthermore, as summarised in Fig. 9.7, there is a qualitative difference in the singlet fission mechanism in non-polar versus intermediate- and highly-polar solvents, while the final triplet yield also depends strongly on the excitation energy. These properties are all uncommon among singlet fission systems. It is therefore important to incorporate vibrational and solvent effects into our model description of DT-Mes, as well as the role of excess energy.

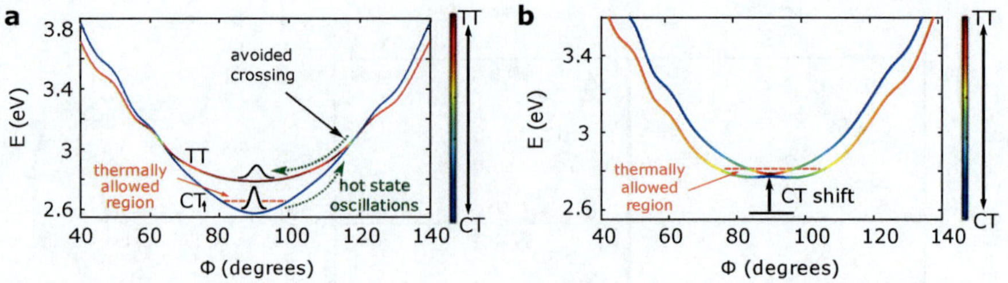

Fig. 9.8 **a** Potential energy surfaces in non-polar solvents (e.g. hexane). $CT_{\uparrow}$ does not mix with TT (colour scale) in the thermally allowed region at room temperature, the boundaries of which are indicated with the red dashed line. Incoherent singlet fission takes place when enough excess energy to reach the avoided crossing to the TT surface is provided. **b** Potential energy surfaces for intermediate-polarity solvents (e.g. ethanol). $CT_{\uparrow}$ is destabilised (arrow) and approaches TT, and the two are strongly mixed by the dihedral oscillation (colour scale), resulting in coherent singlet fission. Exciting $CT_{\uparrow}$ with more excess energy leads to superposition states with stronger TT contributions than within the thermally allowed region of $CT_{\uparrow}$ (red dashed line)

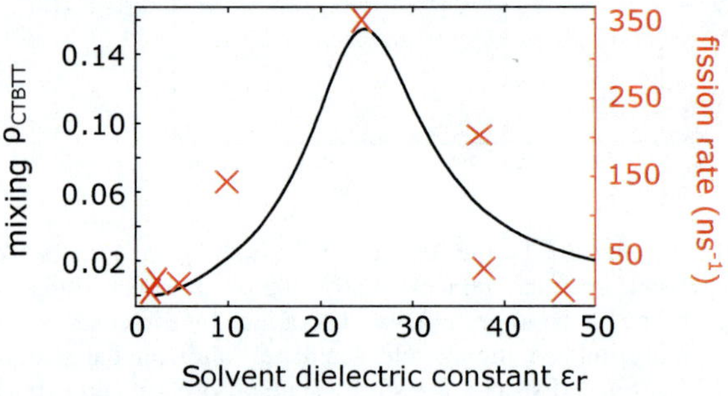

Fig. 9.9 Mixing between bright CT and TT as a function of solvent ϵ_r, at the fixed angle of $\Phi = 80°$. Intermediate-polarity solvents result into maximal mixing and coherent singlet fission. This is compared with the experimental singlet fission rate and the TT yields in Fig. 9.7. The rate is optimised in solvents where singlet fission is coherent

The approximate C_2 symmetry of DT-Mes places important constraints on the fission mechanism. The TT state, which we calculate as the ground state quintet ^{5}TT, is A-symmetric. So is the Φ rotation of Fig. 9.1a, meaning it only mixes states of the same symmetry. Therefore, neglecting two-electron contributions [2, 30, 32], TT only mixes with CT_A along this coordinate. However, only the bright B-symmetry states LE_B and CT_B are optically accessible. The transition from these states to TT must thus be accomplished through some form of symmetry breaking. Physically, such symmetry breaking could be provided by the vibronic coupling along a B-symmetric mode. We take $\eta = 20$ meV as a representative maximum value for this vibronic coupling, similar to values obtained for inter-molecular symmetry-breaking modes in rubrene [10]. We introduce η in our model as the coupling of A- and B-symmetry states to each other. Therefore, a model Hamiltonian which includes the effect of the torsion and symmetry-breaking mode on the electronic states, in the basis of $\{\mathrm{LE}_B, \mathrm{LE}_A, \mathrm{CT}_B, \mathrm{CT}_A, \mathrm{TT}\}$ is:

$$H_{el} = \begin{bmatrix} E_{\mathrm{LE}_B}(\phi) & \eta & J(\phi) & 0 & 0 \\ \eta & E_{\mathrm{LE}_A}(\phi) & 0 & 0 & 0 \\ J(\phi) & 0 & E_{\mathrm{CT}_B}(\phi) & \eta & 0 \\ 0 & 0 & \eta & E_{\mathrm{CT}_A}(\phi) & \Lambda(\phi) \\ 0 & 0 & 0 & \Lambda(\phi) & E_{\mathrm{TT}}(\phi) \end{bmatrix}. \tag{9.6}$$

Here $J(\phi)$ and $\Lambda(\phi)$ are the couplings between states of the same symmetry along the symmetric torsional mode. We optimise the molecular structure along the angle ϕ using DFT to account for the steric effects, and then compute the singlet and TT energies at all intermediate points. By writing the singlet adiabatic states in the electronic basis that we have defined, we obtain the coupling $J(\phi)$.

Regarding the one-electron coupling of TT to the CT states [30]:

$$\langle \mathrm{CT}_{m\to n}| \hat{H}_{el} |\mathrm{TT}\rangle = \sqrt{\frac{3}{2}} \langle l_n| \hat{F} |h_m\rangle , \tag{9.7}$$

where h_m is the HOMO of monomer m, l_n the LUMO of monomer n and $\hat{F}$ the Fock operator. In this case, we work in the basis of $\mathrm{CT}_{A,B}$ which are linear combinations of $\mathrm{CT}_{m\to n}$ and $\mathrm{CT}_{n\to m}$, meaning that the coupling to CT_A will be $\Lambda(\phi) = \sqrt{2}\sqrt{\frac{3}{2}} \langle l_n| \hat{F} |h_m\rangle = \sqrt{3}t_{lh}$, and the one to CT_B will vanish. At this point we would also like to point out that this coupling element refers to that between CT_A and the actual singlet configuration of the TT state, as this was written down in the basis of the four frontier molecular orbitals in Ref. [30]. Therefore, the quintet approximation only affects the exact energy of the TT state (by $\approx$ 10 meV) and not the coupling elements which appear in the Hamiltonian.

The value of the transfer integral t_{lh} can be calculated as $C_{l\mu}C_{h\nu}\beta\cos(\phi)$, with $C_{l\mu}$ and $C_{h\nu}$ being the monomer LUMO and HOMO coefficients on the atoms μ and ν which define the covalent bond between the two monomers. These coefficients are obtained from PPP calculations, while $\beta = 2.2$ eV a parameter taken from the literature for the coupling strength over a single bond [73], and ϕ the dihedral angle between the tetracene monomers.

Due to the negligible Huang–Rhys factors of antisymmetric B modes compared to symmetric ones, their Franck–Condon factors and displacements are orders of magnitude smaller than those of A modes [74–76]. Indeed we found in Chap. 8 that symmetry-breaking 'coupling' modes have negligible displacements compared to the symmetric 'tuning modes'. Therefore, within our model approximation, we consider excess energy to only result in the displacement of A-symmetry modes, and to not generate any additional B symmetry vibrations. Since the singlet fission timescale in DT-Mes is relatively long ($>$ 10 ps), we expect high-frequency A modes displaced via excess energy excitation to have mostly relaxed towards lower-frequency modes at these timescales. The results of Chap. 7 support this assumption, as we found relaxation timescales of less than 4 ps for high-frequency modes. This value of 4 ps is even an overestimate of the actual relaxation timescale due to the absence of explicit anharmonicity in the computational model used in Chap. 7. We further simplify the problem by approximating the resulting low-frequency mode distribution only in terms of motion along Φ. Consequently, within this model approach, excess-energy excitation leads to larger-amplitude oscillations along Φ, accessing higher regions of that surface. Importantly for this picture, the initial step of vibrational relaxation does not remove energy from the molecule but only redistributes it into other motions. These are damped through subsequent vibrational cooling, which for isolated molecules in solution is mediated by solvent-solute interactions, e.g. collisions between DT-Mes and solvent molecules. This diffusion-mediated cooling can require many tens of ps [77–80]. The excess vibrational energy in this 'hot' state manifested as large-amplitude Φ oscillations can thus persist on timescales relevant to singlet fission.

We now turn our attention to incorporating solvent effects into our description. In a polar solvent, local electric fields may be randomly oriented, leading to an energetic separation of the two CT states which inevitably appear in such a dimer molecule, with dipoles pointing in opposite directions. We refer to the stabilised and destabilised CT states as $CT_\downarrow$ and $CT_\uparrow$. These states are symmetry-broken, as the solvent can generally arrange itself in a non-symmetric fashion around DT-Mes. This phenomenon is mathematically captured by introducing an additional coupling Δ between symmetry-pure CT_A and CT_B, which mixes them into the new eigenstates:

$$H_{el} = \begin{bmatrix} E_{\mathrm{LE}_B}(\phi) & \eta & J(\phi) & 0 & 0 \\ \eta & E_{\mathrm{LE}_A}(\phi) & 0 & 0 & 0 \\ J(\phi) & 0 & E_{\mathrm{CT}_B}(\phi) & \eta+\Delta & 0 \\ 0 & 0 & \eta+\Delta & E_{\mathrm{CT}_A}(\phi) & \Lambda(\phi) \\ 0 & 0 & 0 & \Lambda(\phi) & E_{\mathrm{TT}}(\phi) \end{bmatrix}. \tag{9.8}$$

Physically, this expresses the fact that in the presence of an electric field it is the CT states with a permanent dipole and not the symmetry-adapted CT states which are eigenstates of the system.

We can determine the values of Δ corresponding to different solvent ϵ_r by benchmarking against the measured energetic stabilisation $CT_\downarrow$ (Fig. 9.7b). For intermediate-polarity solvents, we find that $CT_\downarrow$ and LE_B are energetically very close, leading to a strong mixing and transfer of oscillator strength from LE to CT, i.e. intensity borrowing [81]. This description closely reproduces the experimental trend of Fig. 9.1c for the molar extinction coefficient, which is minimised in intermediate-polarity solvents. The calculated molar extinction coefficient is shown in Fig. 9.10 for a range of solvents, with the method of obtaining this result outlined in Appendix 1.3.

If the equivalently destabilised state $CT_\uparrow$ were to form, we would in principle expect it to relax into $CT_\downarrow$. However, this process should occur through reorganisation of the solvent shell and would be expected to occur on the few- to tens-picosecond timescale [82]. Accordingly, we would not expect to detect any photon emission

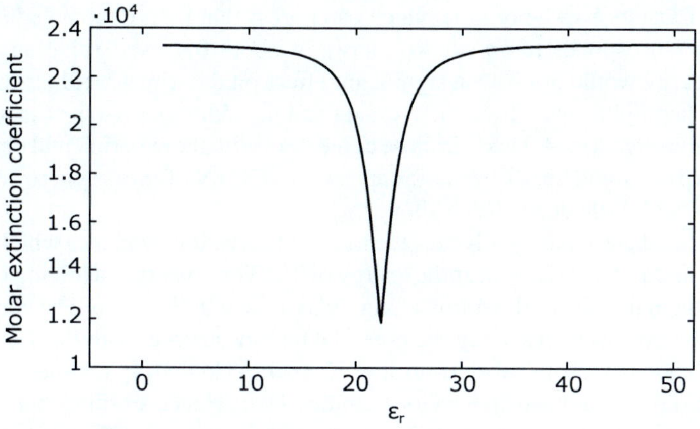

Fig. 9.10 Molar extinction coefficient of DT-Mes in different solvents, calculated within the model approach of the previous sections

from $CT_{\uparrow}$, since the relevant emission lifetime is tens of nanoseconds (for $CT_{\downarrow}$). This timescale remains sufficiently long, though, for $CT_{\uparrow}$ to play an important role in singlet fission despite being an unstable state.

9.8 Coherent and Incoherent Singlet Fission

We can use this framework for the DT-Mes electronic structure to rationalise the surprising behaviour in Fig. 9.7. The transient absorption experiments reveal TT is never formed directly from initial LE_B but is always preceded by CT states (Fig. 9.5b), and it only forms in conditions where CT is already formed with high efficiency (Fig. 9.7a). The only major distinction between fission regimes occurs in the subsequent transition, from CT to TT, where the symmetry of the system must be broken. This step is the chief focus of our analysis, and we consider two limiting cases. In low-polarity solvents, $\eta > \Delta$ and vibronic effects provide the symmetry-breaking needed to access TT. In intermediate-polarity solvents, $\eta << \Delta$ and the symmetry-breaking is dominated by solvent effects. In either case, the CT states formed may initially be vibrationally 'hot' due to excess-energy excitation. Moreover, potentially either $CT_{\downarrow}$ or $CT_{\uparrow}$ may be present, since both are mixed with CT_B and thus accessible from LE_B.

The $CT_{\uparrow}$ and TT potential energy surfaces along Φ for low-polarity solvents are given in Fig. 9.8a, coded for their CT/TT character. In this case, $CT_{\uparrow}$ and $CT_{\downarrow}$ are almost degenerate. $CT_{\uparrow}$ does not develop any TT character within the region accessible at room temperature (nor does $CT_{\downarrow}$), and the TT surface also remains pure. Within this model, a transition between the states is possible only through the avoided crossing that appears at larger angles, making singlet fission incoherent

[10, 19]. Excitation of a hot state which can access that part of the potential energy landscape corresponds to an excess energy of about 0.4 eV. Excitation with still greater energy would not have a significant effect on the triplet yield, as the mixing is always negligible away from the crossing and no additional B-type (i.e. coupling) vibrations are generated. This is in close agreement with the experimental observation that a sharp threshold for TT formation appears at 0.33 eV of excess-energy excitation (Fig. 9.7), with little dependence above that.

In intermediate-polarity solvents, the transient solvation dynamics which stabilise $CT_{\downarrow}$ also destabilise $CT_{\uparrow}$ to near the energy of TT. This results in a mixing of the two states through the dihedral rotation along Φ, as shown in Fig. 9.8b. The contribution of each component varies along the potential energy surface, with the TT character (red shading) increasing away from the FC point. This is the regime of coherent singlet fission, as defined in previous studies [10]. Hence exciting the system at higher energies leads to superposition states which are more 'TT rich', leading to stronger couplings to the final, relaxed TT state. A superposition of TT and $CT_{\uparrow}$ eventually dephases towards its constituents, which are the relevant eigenstates of the system at its equilibrium geometry. The larger the TT contribution, the more likely it is that the system will collapse towards the final, relaxed TT state. In this regime the destabilised $CT_{\uparrow}$ state is the 'gateway' for singlet fission regardless of excitation energy, and we would thus expect the TT yield to track the accessibility of $CT_{\uparrow}$. Experimentally, this would translate into the energy distribution inferred in Fig. 9.7a (right), which is exactly what is observed. At the same time, regions of the surface with high TT contributions exhibit reduced CT character, leading to an anti-correlation of the TT and long-lived CT yields.

The increased degree of mixing between the CT and TT surfaces with increasing solvent polarity is reflected in the experimental fission rates shown in Fig. 9.9 and the TT yields in Fig. 9.7a. Initially, increasing the solvent polarity increases the rate of singlet fission, but in very polar solvents the TT formation rate drops again to low values. In these solvents $CT_{\uparrow}$ shifts above the TT surface and the two stop mixing, leading to a regime of incoherent fission similar to the non-polar case. This mixing is quantified via $\rho_{\mathrm{CT}_B\mathrm{TT}}$ of Eq. (9.5), which we plot in Fig. 9.9 for the fixed angle $\Phi = 80^\circ$ as a function of solvent dielectric constant. Note that this is calculated in the original basis of symmetry-pure CT states which form $CT_{\uparrow}$. In summary, we find that intermediate-polarity solvents lead to maximal mixing between CT_B and TT, thus leading to coherent singlet fission. In this case, TT is formed via the dephasing of the $CT_{\uparrow}$/TT superposition, which is more efficient and has qualitative differences compared to incoherent TT formation through an avoided crossing. Hence intermediate polarity solvents provide an optimal regime for singlet fission, qualitatively reproducing the experimental trend for the singlet fission rates. The different singlet fission regimes are summarised in the multidimensional plot of Fig. 9.11 for the $CT_{\uparrow}$ and TT surfaces. For the different dihedral angles and solvent dielectric constants, the surfaces are annotated with the respective CT/TT contribution as a colour code.

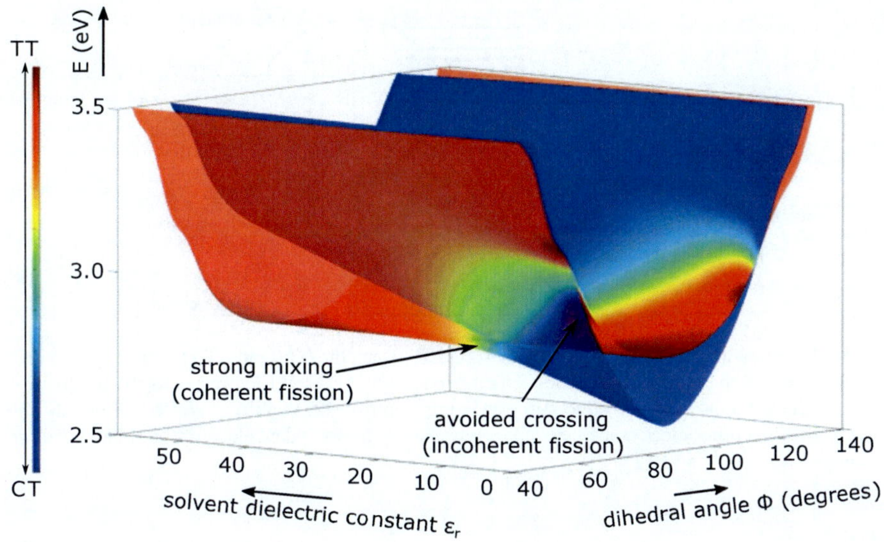

Fig. 9.11 Adiabatic potential energy surfaces along the dihedral angle Φ and the solvent dielectric constant, showing their CT (blue) and TT (red) character. For low-polarity solvents the surfaces only mix for large-amplitude oscillations along Φ in the vicinity of an avoided crossing, leading to incoherent fission. Strong mixing is induced in intermediate-polarity solvents (e.g. ethanol), making singlet fission coherent. Finally, very polar solvents once again lead to reduced mixing between the surfaces

9.9 Conclusions

Our results demonstrate not only that intramolecular singlet fission in DT-Mes is a 'hot' process mediated by an unusual 'destabilised' CT state, but also that the system can explore coherent and incoherent regimes of singlet fission. This is achieved by exploiting symmetry-breaking solvent interactions to induce the necessary energetic proximity for a vibrational mode to mix the CT and TT states. Interestingly, these interactions are by their nature transient, driven by changes in the solvation shell. Here, the precise degree of mixing depends on the solvent polarity, allowing us to switch between coherent and incoherent fission in different solvents and explore the effects in detail through altering the energy of excitation. In addition, while coherent singlet fission has been reported to occur due to mixing between the bright singlet and TT states [7, 8, 10, 19], it is to our knowledge the first time that the role of CT/TT mixing for coherent fission is studied in detail. The fact that singlet fission in DT-Mes occurs through a real CT intermediate underlines the importance of this second step of fission and the advantages of a system dependent on CT states, which are markedly easier to control through environmental factors such as solvent polarity than LE states.

Our approach for controlling the fission mechanism also indicates a more general concept, where various symmetry-breaking effects could be used as 'switches'

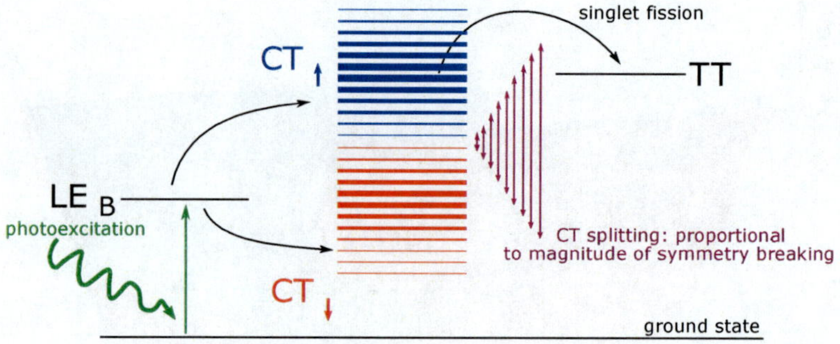

Fig. 9.12 Following photoexcitation to the bright LE_B state, the $CT_\uparrow$ and $CT_\downarrow$ states get populated. The thickness of these two states denotes their mixing with TT and LE_B respectively. Symmetry-breaking interactions (solvent effects in our study) control their splitting and consequently the mixing with the other electronic states. Coherent singlet fission takes place in the regime of high mixing between $CT_\uparrow$ and TT

between coherent and incoherent regimes. External electric fields and strong coupling to light [83] could potentially provide a similar symmetry-breaking effect on the CT states, leading to a coherent fission regime in the vicinity of large mixing, as visualised in Fig. 9.12. The same could be achieved through chemical synthesis, or through changes in the crystal symmetry. Experiments [11] and calculations [84] indicate that the rate of fission closely depends on the crystal structure. Tetracene and TIPS-tetracene exhibit strikingly different fission properties for highly crystalline versus symmetry-broken amorphous or polycrystalline films [11, 17, 85], indicating a possible transition between coherent and incoherent fission within the same material. Singlet fission is but one example of a photophysical process where the efficiency may be tuned by entering a coherent regime. The same underlying principle could also be used to manipulate and ultimately enhance processes as diverse as long-range energy transport, biomimetic light harvesting and charge separation at interfaces in solar cells [86, 87].

Appendix 1: Experimental Details

While the experimental work presented in this chapter was not undertaken by the author of this thesis, the relevant details are given here for completeness.

1.1 Sample Preparation

The molecular mechanism of singlet fission was studied using solutions of DT-Mes in solvents of varying polarity. The mesityl substituents provide chemical stability and render limited solubility. Unless otherwise noted, all measurements in this work were performed on solutions with concentration of 0.5 mg/mL, prepared and sealed under nitrogen atmosphere. Triplet sensitisation for transient-absorption measurements was performed using established procedures [47], with a mixed solution of 0.5 mg/mL DT-Mes and 1.5 mg/mL N-methylfulleropyrrolidine (NMFP) in toluene. All solution measurements were performed in 1 mm light-path quartz cuvettes (Hellma Analytics). To prepare highly viscous solutions, stock solutions of 1 mg/mL DT-Mes and 20 mg/mL polystyrene, in toluene were mixed together to obtain a final DT-Mes concentration of 0.17 mg/mL with a DT-Mes:polymer weight ratio of 1 : 99. To obtain samples that gave phosphorescence, DT-Mes was mixed with platinum octaethylporphyrin (PtOEP, purchased from Sigma Aldrich) and dispersed in polystyrene matrix. The final weight percentage of DT-Mes: PtOEP: polymer in the mixture was 1 : 4 : 95. This mixture was drop-casted on Spectrosil® quartz substrates in nitrogen atmosphere.

Chemical Reduction-Oxidation. Chemical doping was used to generate radical cation and anion (D+A-) in DT-Mes (Fig. 9.5d). $SbCl_5$ and Li in THF were used as oxidising and reducing agents, respectively.

1.2 Experimental Setup

Photoluminescence Spectroscopy. The PL was measured in two distinct temporal regimes. Fast (sub-100-ps) photoluminescence dynamics were studied using the transient grating technique described in detail in Chen et al. [60], excited at ≈ 500 nm. Longer-time dynamics were recorded with a standard time-correlated single-photon counting system (Edinburgh Instruments), using 40 MHz excitation at 474 nm (PicoQuant). Phosphorescence was detected using a calibrated infrared InGaAs photodiode array (ANDOR iDus 490A) coupled to a spectrograph (ANDOR Shamrock), with CW excitation at 532 nm (1.1 mW).

Transient Absorption Spectroscopy. Transient absorption measurements were performed on a previously reported setup [47]. Briefly, broad-band probe pulses were generated using noncollinear optical parametric amplifiers (NOPAs) built in-house to cover two separate spectral ranges: 500–800 nm, and 800–1150 nm. The same InGaAs array detector (Hamamatsu G11608-512) was used for all wavelengths. For sub-picosecond resonant excitation, DT-Mes was pumped with the output from an automated OPA (TOPAS, Light Conversion), with a pulse duration of < 200 fs, unless otherwise mentioned. The sub-ps setup was limited by the length of the mechanical delay stage to delays of ≈ 2 ns. Further spectral evolution was investigated using excitation with the ≈ 3 ns output of a Nd-YAG laser (Ekspla). Triplet sensitisation was investigated using excitation with the ≈ 1 ns output of a frequency-doubled (532 nm) Q-switched Nd-YVO4 laser (Advanced Optical Technologies), which was externally triggered with an electronic pulse. For these measurements, strong pump scatter in the spectral range 520–540 nm required removal of this probe region. In all measurements, pump and probe polarisations were set to magic angle (54.7°). Typical excitation densities were $10^{14} - 10^{15}$ photons/pulse/cm^2, and all decay kinetics were found to be independent of pump intensity.

Photoluminescence Quantum Efficiency. The PL quantum efficiency was determined on the same system as used for phosphorescence measurement versus Nile Blue reference ($\lambda_{exc} = 540\,nm$, QY = 0.27) [88].

1.3 Calculation of the Molar Extinction Coefficient

We calculate the molar extinction coefficient from the oscillator strength as described in Chap. 4 of Ref. [89]. The absorption spectrum of DT-Mes extends from approximately 350 to 520 nm, giving us the relationship:

$$\epsilon = 0.267 * 10^5 * f \ (L \cdot mol^{-1} \cdot cm^{-1}) \quad (9.9)$$

This theoretical trend for the molar extinction coefficient qualitatively reproduces the experimental one of Fig. 9.1c, supporting the proposal that changes in the extinction coefficient are due to the mixing of bright LE_B with dark CT states. However, the 'dip' in the calculated extinction coefficient is much sharper compared to the experimentally measured one. We believe this to be due to the infinitesimal width of the electronic states within the PPP framework—the states only mix efficiently when they are very close to resonance, therefore the extinction coefficient only decreases within a very narrow region of solvent dielectric constants. In reality, both the LE_B and $CT_\downarrow$ have a finite width due to static and dynamic disorder, allowing them to mix more efficiently in a wider range of solvents, leading to a more gradual decrease (and then increase) of the extinction coefficient as observed in Fig. 9.1c.

References

1. Alvertis AM, Lukman S, Hele TJH, Fuemmeler EG, Feng J, Wu J, Greenham NC, Chin AW, Musser AJ (2019) Switching between coherent and incoherent singlet fission via solvent-induced symmetry breaking. J Am Chem Soc 141:17558
2. Smith MB, Michl J (2010) Singlet fission. Chem Rev 110:6891
3. Hanna MC, Nozik AJ (2006) Solar conversion efficiency of photovoltaic and photoelectrolysis cells with carrier multiplication absorbers. J Appl Phys 100:074510
4. Rao A, Friend RH (2017) Harnessing singlet exciton fission to break the Shockley–Queisser limit. Nat Rev Mater 2:17063
5. Congreve DN, Lee J, Thompson NJ, Hontz E (2013) External quantum efficiency above 100% in a singlet-exciton-fission-based organic photovoltaic cell. Science 340:334
6. Chan WL, Ligges M, Jailaubekov AE, Kaake LG, Miaja-Avila L, Zhu X-Y (2011) Observing the multiexciton state in singlet fission and ensuing ultrafast multielectron transfer. Science 334:1541

7. Chan WL, Berkelbach TC, Provorse MR, Monahan NR, Tritsch JR, Hybertsen MS, Reichman DR, Gao J, Zhu XY (2013) The quantum coherent mechanism for singlet fission: experiment and theory. Acc Chem Res 46:1321
8. Greyson EC, Vura-Weis J, Michl J, Ratner MA (2010) Maximizing singlet fission in organic dimers: theoretical investigation of triplet yield in the regime of localized excitation and fast coherent electron transfer. J Phys Chem B 114:14168
9. Monahan NR, Sun D, Tamura H, Williams KW, Xu B, Zhong Y, Kumar B, Nuckolls C, Harutyunyan AR, Chen G, Dai HL, Beljonne D, Rao Y, Zhu XY (2017) Dynamics of the triplet-pair state reveals the likely coexistence of coherent and incoherent singlet fission in crystalline hexacene. Nat Chem 9:341
10. Miyata K, Kurashige Y, Watanabe K, Sugimoto T, Takahashi S, Tanaka S, Takeya J, Yanai T, Matsumoto Y (2017) Coherent singlet fission activated by symmetry breaking. Nat Chem 9:983
11. Yost SR, Lee J, Wilson MWB, Wu T, McMahon DP, Parkhurst RR, Thompson NJ, Congreve DN, Rao A, Johnson K, Sfeir MY, Bawendi MG, Swager TM, Friend RH, Baldo M, Van Voorhis T (2014) A transferable model for singlet-fission kinetics. Nat Chem 6:492
12. Zimmerman PM, Zhang Z, Musgrave CB (2010) Singlet fission in pentacene through multi-exciton quantum states. Nat Chem 2:648
13. Swenberg CE, Stacy WT (1968) Bimolecular radiationless transitions in crystalline tetracene. Chem Phys Lett 2:327
14. Burdett JJ, Bardeen CJ (2012) Quantum beats in crystalline tetracene delayed fluorescence due to triplet pair coherences produced by direct singlet fission. J Am Chem Soc 134:8597
15. Wilson MWB, Rao A, Johnson K, Gélinas S, Di Pietro R, Clark J, Friend RH (2013) Temperature-independent singlet exciton fission in tetracene. J Am Chem Soc 135:16680
16. Tayebjee MJY, Clady RGCR, Schmidt TW (2013) The exciton dynamics in tetracene thin films. Phys Chem Chem Phys 15:14797
17. Stern HL, Cheminal A, Yost SR, Broch K, Bayliss SL, Chen K, Tabachnyk M, Thorley K, Greenham N, Hodgkiss JM, Anthony J, Head-Gordon M, Musser AJ, Rao A, Friend RH (2017) Vibronically coherent ultrafast triplet-pair formation and subsequent thermally activated dissociation control efficient endothermic singlet fission. Nat Chem 9:1205
18. Burdett JJ, Gosztola D, Bardeen CJ (2011) The dependence of singlet exciton relaxation on excitation density and temperature in polycrystalline tetracene thin films: kinetic evidence for a dark intermediate state and implications for singlet fission. J Chem Phys 135:214508
19. Chan WL, Ligges M, Zhu XY (2012) The energy barrier in singlet fission can be overcome through coherent coupling and entropic gain. Nat Chem 4:840
20. Musser AJ, Liebel M, Schnedermann C, Wende T, Kehoe TB, Rao A, Kukura P (2015) Evidence for conical intersection dynamics mediating ultrafast singlet exciton fission. Nat Phys 11:352
21. Bakulin AA, Morgan SE, Kehoe TB, Wilson MWB, Chin AW, Zigmantas D, Egorova D, Rao A (2016) Real-time observation of multiexcitonic states in ultrafast singlet fission using coherent 2D electronic spectroscopy. Nat Chem 8:16
22. Tempelaar R, Reichman DR (2018) Vibronic exciton theory of singlet fission. III. How vibronic coupling and thermodynamics promote rapid triplet generation in pentacene crystals. J Chem Phys 148:244701
23. Fujihashi Y, Chen L, Ishizaki A, Wang J, Zhao Y (2017) Effect of high-frequency modes on singlet fission dynamics. J Chem Phys 146:044101
24. Tamura H, Huix-Rotllant M, Burghardt I, Olivier Y, Beljonne D (2015) First-principles quantum dynamics of singlet fission: coherent versus thermally activated mechanisms governed by molecular π stacking. Phys Rev Lett 115:107401
25. Feng X, Krylov AI (2016) On couplings and excimers: lessons from studies of singlet fission in covalently linked tetracene dimers. Phys Chem Chem Phys 18:7751
26. Huang Z, Fujihashi Y, Zhao Y (2017) Effect of off-diagonal exciton-phonon coupling on intramolecular singlet fission. J Phys Chem Lett 8:3306

27. Schröder FAYN, Turban DHP, Musser AJ, Hine NDM, Chin AW (2019) Tensor network simulation of multi-environmental open quantum dynamics via machine learning and entanglement renormalisation. Nat Commun 10:1062
28. Lukman S, Chen K, Hodgkiss JM, Turban DHP, Hine NDM, Dong S, Wu J, Greenham NC, Musser AJ (2016) Tuning the role of charge-transfer states in intramolecular singlet exciton fission through side-group engineering. Nat Commun 7:13622
29. Fuemmeler EG, Sanders SN, Pun AB, Kumarasamy E, Zeng T, Miyata K, Steigerwald ML, Zhu XY, Sfeir MY, Campos LM, Ananth N (2016) A direct mechanism of ultrafast intramolecular singlet fission in pentacene dimers. ACS Cent Sci 2:316
30. Berkelbach TC, Hybertsen MS, Reichman DR (2013) Microscopic theory of singlet exciton fission. II. Application to pentacene dimers and the role of superexchange. J Chem Phys 138:114103
31. Zimmerman PM, Bell F, Casanova D, Head-Gordon M (2011) Mechanism for singlet fission in pentacene and tetracene: from single exciton to two triplets. J Am Chem Soc 133:19944
32. Smith MB, Michl J (2013) Recent advances in singlet fission. Annu Rev Phys Chem 64:361
33. Berkelbach TC, Hybertsen MS, Reichman DR (2013) Microscopic theory of singlet exciton fission. I. General formulation. J Chem Phys 138:114102
34. Krishnapriya KC, Roy P, Puttaraju B, Salzner U, Musser AJ, Jain M, Dasgupta J, Patil S (2019) Spin density encodes intramolecular singlet exciton fission in pentacene dimers. Nat Commun 10:33
35. Sanders SN, Kumarasamy E, Pun AB, Trinh MT, Choi B, Xia J, Taffet EJ, Low JZ, Miller JR, Roy X, Zhu XY, Steigerwald ML, Sfeir MY, Campos LM (2015) Quantitative intramolecular singlet fission in bipentacenes. J Am Chem Soc 137:8965
36. Zirzlmeier J, Lehnherr D, Coto PB, Chernick ET, Casillas R, Basel BS, Thoss M, Tykwinski RR, Guldi DM (2015) Singlet fission in pentacene dimers. Proc Natl Acad Sci 112:5325
37. Korovina NV, Das S, Nett Z, Feng X, Joy J, Haiges R, Krylov AI, Bradforth SE, Thompson ME (2016) Singlet fission in a covalently linked cofacial alkynyltetracene dimer. J Am Chem Soc 138:617
38. Johnson JC, Nozik AJ, Michl J (2013) The role of chromophore coupling in singlet fission. Acc Chem Res 46:1290
39. Kumarasamy E, Sanders SN, Tayebjee MJY, Asadpoordarvish A, Hele TJH, Fuemmeler EG, Pun AB, Yablon LM, Low JZ, Paley DW, Dean JC, Choi B, Scholes GD, Steigerwald ML, Ananth N, McCamey DR, Sfeir MY, Campos LM (2017) Tuning singlet fission in π-bridge-π chromophores. J Am Chem Soc 139:12488
40. Basel BS, Zirzlmeier J, Hetzer C, Phelan BT, Krzyaniak MD, Reddy SR, Coto PB, Horwitz NE, Young RM, White FJ, Hampel F, Clark T, Thoss M, Tykwinski RR, Wasielewski MR, Guldi DM (2017) Unified model for singlet fission within a non-conjugated covalent pentacene dimer. Nat Commun 8:15171
41. Cook JD, Carey TJ, Arias DH, Johnson JC, Damrauer NH (2017) Solvent-controlled branching of localized versus delocalized singlet exciton states and equilibration with charge transfer in a structurally well-defined tetracene dimer. J Phys Chem A 121:9229
42. Korovina NV, Joy J, Feng X, Feltenberger C, Krylov AI, Bradforth SE, Thompson ME (2018) Linker-dependent singlet fission in tetracene dimers. J Am Chem Soc 140:10179
43. Yamakado T, Takahashi S, Watanabe K, Matsumoto Y, Osuka A, Saito S (2018) Conformational planarization versus singlet fission: distinct excited-state dynamics of cyclooctatetraene-fused acene dimers. Angew Chem Int Ed 57:5438
44. Sakuma T, Sakai H, Araki Y, Mori T, Wada T, Tkachenko NV, Hasobe T (2016) Long-lived triplet excited states of bent-shaped pentacene dimers by intramolecular singlet fission. J Phys Chem A 120:1867
45. Lukman S, Musser AJ, Chen K, Athanasopoulos S, Yong CK, Zeng Z, Ye Q, Chi C, Hodgkiss JM, Jishan W, Friend RH, Greenham NC (2015) Tuneable singlet exciton fission and triplet-triplet annihilation in an orthogonal pentacene dimer. Adv Funct Mater 25:5452
46. Margulies EA, Miller CE, Yilei W, Ma L, Schatz GC, Young RM, Wasielewski MR (2016) Enabling singlet fission by controlling intramolecular charge transfer in π-stacked covalent terrylenediimide dimers. Nat Chem 8:1120

47. Musser AJ, Al-Hashimi M, Maiuri M, Brida D, Heeney M, Cerullo G, Friend RH, Clark J (2013) Activated singlet exciton fission in a semiconducting polymer. J Am Chem Soc 135:12747
48. Elich K, Kitazawa M, Okada T, Wortmann R (1997) Effect of S1 torsional dynamics on the time-resolved fluorescence spectra of 9,9-bianthryl in solution. J Phys Chem A 101:2010
49. Mo Y, Lei J, Sun Y, Zhang Q, Wei G (2016) Conformational ensemble of hIAPP dimer: insight into the molecular mechanism by which a green tea extract inhibits hIAPP aggregation. Sci Rep 6:33076
50. Coulson CA, Rushbrooke GS (1940) Note on the method of molecular orbitals. Math Proc Camb Philos Soc 36:193
51. Hestand NJ, Yamagata H, Xu B, Sun D, Zhong Y, Harutyunyan AR, Chen G, Dai HL, Rao Y, Spano FC (2015) Polarized absorption in crystalline pentacene: theory vs experiment. J Phys Chem C 119:22137
52. Pople JA (1953) Electron interaction in unsaturated hydrocarbons. Trans Faraday Soc 49:1375
53. Pariser R, Parr RG (1953) A semi-empirical theory of the electronic spectra and electronic structure of complex unsaturated molecules. I. J Chem Phys 21:466
54. Robinson GW (1967) Intensity enhancement of forbidden electronic transitions by weak intermolecular interactions. J Chem Phys 46:572
55. Petrozza A, Laquai F, Howard IA, Kim J-S, Friend RH (2010) Dielectric switching of the nature of excited singlet state in a donor-acceptor-type polyfluorene copolymer. Phys Rev B 81:205421
56. Fukuzumi S, Kotani H, Ohkubo K, Ogo S, Tkachenko NV, Lemmetyinen H (2004) Electron-transfer state of 9-mesityl-10-methylacridinium ion with a much longer lifetime and higher energy than that of the natural photosynthetic reaction center. J Am Chem Soc 126:1600
57. Eaton SW, Shoer LE, Karlen SD, Dyar SM, Margulies EA, Veldkamp BS, Ramanan C, Hartzler DA, Savikhin S, Marks TJ, Wasielewski MR (2013) Singlet exciton fission in polycrystalline thin films of a slip-stacked perylenediimide. J Am Chem Soc 135:14701
58. Stern HL, Musser AJ, Gelinas S, Parkinson P, Herz LM, Bruzek MJ, Anthony J, Friend RH, Walker BJ (2015) Identification of a triplet pair intermediate in singlet exciton fission in solution. Proc Natl Acad Sci 112:7656
59. Nagashima H, Kawaoka S, Akimoto S, Tachikawa T, Matsui Y, Ikeda H, Kobori Y (2018) Singlet-fission-born quintet state: sublevel selections and trapping by multiexciton thermodynamics. J Phys Chem Lett 9:5855
60. Chen K, Gallaher JK, Barker AJ, Hodgkiss JM (2014) Transient grating photoluminescence spectroscopy: an ultrafast method of gating broadband spectra. J Phys Chem Lett 5:1732
61. Jurczok M, Plaza P, Martin MM, Meyer YH, Rettig W (2000) Excited state relaxation paths in 9,9'-bianthryl and 9-carbazolyl-anthracene: a sub-ps transient absorption study. Chem Phys 253:339
62. Burgdorff C, Kircher T, Löhmannsröben HG (1988) Photophysical properties of tetracene derivatives in solution. Spectrochim Acta Part A Mol Spectrosc 44:1137
63. Rao A, Chow PCY, Gélinas S, Schlenker CW, Li C-Z, Yip H-L, Jen AK-Y, Ginger DS, Friend RH (2013) The role of spin in the kinetic control of recombination in organic photovoltaics. Nature 500:435
64. Sanders SN, Kumarasamy E, Pun AB, Steigerwald ML, Sfeir MY, Campos LM (2016) Intramolecular singlet fission in oligoacene heterodimers. Angew Chem Int Ed 55:3373
65. Tayebjee MJY, Sanders SN, Kumarasamy E, Campos LM, Sfeir MY, McCamey DR (2017) Quintet multiexciton dynamics in singlet fission. Nat Phys 13:182
66. Busby E, Xia J, Wu Q, Low JZ, Song R, Miller JR, Zhu XY, Campos LM, Sfeir MY (2015) A design strategy for intramolecular singlet fission mediated by charge-transfer states in donor-acceptor organic materials. Nat Mater 14:426
67. Kasai Y, Tamai Y, Ohkita H, Benten H, Ito S (2015) Ultrafast singlet fission in a push-pull low-bandgap polymer film. J Am Chem Soc 137:15980
68. Kolomeisky AB, Feng X, Krylov AI (2014) A simple kinetic model for singlet fission: a role of electronic and entropic contributions to macroscopic rates. J Phys Chem C 118:5188

69. Yong CK, Musser AJ, Bayliss SL, Lukman S, Tamura H, Bubnova O, Hallani RK, Meneau A, Resel R, Maruyama M, Hotta S, Herz LM, Beljonne D, Anthony JE, Clark J, Sirringhaus H (2017) The entangled triplet pair state in acene and heteroacene materials. Nat Commun 8:15953
70. Lukman S, Richter JM, Yang L, Pan H, Jishan W, Greenham NC, Musser AJ (2017) Efficient singlet fission and triplet-pair emission in a family of zethrene diradicaloids. J Am Chem Soc 139:18376
71. Musser AJ, Clark J (2019) Triplet-pair states in organic semiconductors. Annu Rev Phys Chem 70:323
72. Feng X, Luzanov AV, Krylov AI (2013) Fission of entangled spins: an electronic structure perspective. J Phys Chem Lett 4:3845
73. Tavan P, Schulten K (1987) Electronic excitations in finite and infinite polyenes. Phys Rev B 36:4337
74. Clark RJH, Dines TJ (1986) Resonance Raman spectroscopy, and its application to inorganic chemistry. Angew Chem Int Ed 25:131
75. Köppel H, Domcke W, Cederbaum LS (1984) Multimode molecular dynamics beyond the Born–Oppenheimer approximation. Adv Chem Phys 57:102
76. Negri F, Orlandi G (2005) Computational photochemistry, chap IV. Elsevier, p 129
77. Tan X, Gustafson TL, Lefumeux C, Burdzinski G, Buntinx G, Poizat O (2002) Solvation dynamics probed by femtosecond transient absorption spectroscopy: vibrational cooling and conformational relaxation in S1 trans-4,4-diphenylstibene. J Phys Chem A 106:3593
78. Weigel A, Ernsting NP (2010) Excited stilbene: intramolecular vibrational redistribution and solvation studied by femtosecond stimulated Raman spectroscopy. J Phys Chem B 114:7879
79. Grubb MP, Coulter PM, Marroux HJB, Hornung B, McMullen RS, Orr-Ewing AJ, Ashfold MNR (2016) Translational, rotational and vibrational relaxation dynamics of a solute molecule in a non-interacting solvent. Nat Chem 8:1042
80. Grubb MP, Coulter PM, Marroux HJB, Orr-Ewing AJ, Ashfold MNR (2017) Unravelling the mechanisms of vibrational relaxation in solution. Chem Sci 8:3062
81. Hele TJH, Fuemmeler EG, Sanders SN, Kumarasamy E, Sfeir MY, Campos LM, Ananth N (2019) Anticipating acene-based chromophore spectra with molecular orbital arguments. J Phys Chem A 123:2527
82. Kwok WM, Ma C, George MW, Grills DC, Matousek P, Parker AW, Phillips D, Toner WT, Towrie M (2007) Solvent effects on the charge transfer excited states of 4-dimethylaminobenzonitrile (DMABN) and 4-dimethylamino-3,5-dimethylbenzonitrile (TMABN) studied by time-resolved infrared spectroscopy: a direct observation of hydrogen bonding interactions. Photochem Photobiol Sci 6:987
83. Polak D, Jayaprakash R, Lyons TP, Martínez-Martínez L, Leventis A, Fallon KJ, Coulthard H, Bossanyi DG, Georgiou K, Petty AJ, Anthony J, Bronstein H, Yuen-Zhou J, Tartakovskii AI, Clark J, Musser AJ (2020) Manipulating molecules with strong coupling: harvesting triplet excitons in organic exciton microcavities. Chem Sci 11:343
84. Refaely-Abramson S, Da Jornada FH, Louie SG, Neaton JB (2017) Origins of singlet fission in solid pentacene from an ab initio Green's function approach. Phys Rev Lett 119:267401
85. Piland GB, Bardeen CJ (2015) How morphology affects singlet fission in crystalline tetracene. J Phys Chem Lett 6:1841
86. Gélinas S, Rao A, Kumar A, Smith SL, Chin AW, Clark J, van der Poll TS, Bazan GC, Friend RH (2014) Ultrafast long-range charge separation in organic semiconductor photovoltaic diodes. Science 343:512
87. Scholes GD, Fleming GR, Chen LX, Aspuru-Guzik A, Buchleitner A, Coker DF, Engel GS, Van Grondelle R, Ishizaki A, Jonas DM, Lundeen JS, McCusker JK, Mukamel S, Ogilvie JP, Olaya-Castro A, Ratner MA, Spano FC, Whaley KB, Zhu X (2017) Using coherence to enhance function in chemical and biophysical systems. Nature 543:647
88. Sens R, Karl H (1981) Fluorescence quantum yield of oxazine and carbazine laser dyes. J Luminesc 24:709
89. Turro NJ, Ramamurthy V, Scaiano JC (2009) Principles of molecular photochemistry: an introduction. University Science Books, Sausalito, CA

Chapter 10
Conclusions and Outlook

Processes involving excitons in organic semiconductors play a key role in energy harvesting and transfer in these materials. In this thesis, we have examined exciton physics from a theoretical point of view, and in close partnership to experiment. We have shown that in order to gain a thorough understanding of exciton properties one needs to move away from the traditional way of viewing this problem as a purely electronic one, within which interactions with light and vibrations can be considered to be a small perturbation. Our work has demonstrated that these interactions can indeed become extremely strong and has allowed us insights into ways of controlling their magnitude. In turn, this has made it possible in some cases to highlight ways of utilising exciton-photon and exciton-vibration coupling in order to optimise energy harvesting and transfer processes. Below we briefly summarise the important conclusions from the individual chapters, highlighting the potential research avenues that these open in each case.

In Chap. 5 we have developed a first-principles methodology to model the interactions of excitons and photons, describing hybrid so-called exciton-polariton states they form. The highly ordered character of the studied systems, variants of polydiacetylene, was found to result in very strong exciton-photon coupling, manifesting itself in a wide stop-band of the polariton bands. We found that the close-to-perfect dipole alignment of the diacetylene monomers is largely responsible for this, and indeed the stop-band width decreases significantly when we induce a torsion of a mere 14° between subsequent monomers. Furthermore, the dielectric environment was shown to play an important role in determining light-matter interactions, and a larger screening results in a reduction of their strength. Exciton-photon interactions were shown to strongly depend on the conjugation length of polymer chains and hence static disorder could have a significant effect. Finally, we have shown that vibrations also affect exciton-photon interactions. In principle, all of these factors need to be accounted for at the same time and self-consistently, which has thus far remained an open challenge, and future work could move towards this direction.

© The Author(s), under exclusive license to Springer Nature Switzerland AG 2021
A. M. Alvertis, *On Exciton–Vibration and Exciton–Photon Interactions in Organic Semiconductors*, Springer Theses,
https://doi.org/10.1007/978-3-030-85454-6_10

Importantly, while our theoretical framework for studying polaritons requires input on the electronic properties of the studied system, it is agnostic to the exact method used to obtain this. We therefore anticipate that our first-principles approach could be of great practical value to the community, leading to further insights into the control of exciton-photon interactions.

Chapter 6 focuses on the non-perturbative modelling of exciton-vibration coupling in the acene series of organic semiconductors. We found that the spatial delocalisation of excitons is the key parameter controlling the magnitude of this interaction in equilibrium. At 0 K it is more localised excitons that are predominantly affected by vibrations, since nuclear quantum fluctuations of localised high-frequency motions dominate in organic semiconductors. At room temperature on the other hand, delocalised excitons are more strongly affected by thermally-populated phonons that modulate intermolecular distances. By accounting for the effects of nuclear quantum fluctuations on exciton energies, we achieved unprecedented agreement with experiment. We therefore highlighted the importance of these effects and presented the community with a method for accurate exciton energy prediction. We revealed that the temperature dependence of the vast majority of organic semiconductors is weak, due to the competing effects of exciton-phonon coupling and thermal expansion, which individually are both determined by exciton delocalisation. This opens the possibility of controlling exciton energies through temperature for materials with unusual properties, such as negative thermal expansion, which will be the subject of a future work. Finally, we have shown that the pressure dependence of excitons provides a strong indication of their spatial delocalisation and can be used to experimentally probe the magnitude of exciton-phonon interactions in these materials.

In Chap. 7 we shifted our attention to the out-of-equilibrium exciton-vibration dynamics which arise following the photoexcitation of a covalent tetracene dimer system (DT-Mes). In order to study the real-time dynamics of this large system of a pair of excitons in contact with more than one hundred molecular vibrations, we extended upon a previously developed method that is based on tensor networks, which can account for strong exciton-vibration interactions and goes beyond the Born–Oppenheimer approximation that is known to often break down over the course of photoinduced dynamics. We have described the selective excitation of vibrational modes of DT-Mes using excess energy, through which endothermic charge transfer becomes possible. Hence the significant role of vibrations on exciton dynamics became clear. We explored the impact of exciting vibrations of different energies, and found that low-energy vibrations induce more efficient charge transfer due to the limited relaxation pathways available to them, which in turn allows the system to maintain its quantum coherence for longer. To our knowledge, well-controlled selective excitation of vibrational modes of a wide range of energies has yet to be demonstrated in experiments studying photoinduced dynamics. This could therefore be a promising avenue to explore in order to improve the efficiency of light-harvesting processes.

In Chap. 8 we applied our tensor network methodology to the process of exothermic singlet exciton fission in a covalent pentacene dimer (DP-Mes). By rigorously comparing to experiment, we verified the accuracy of our approach, and proceeded to use it in order to map the entirety of real-time exciton-vibration interactions, including those that are invisible to the employed experimental techniques. Hence the microscopic mechanism of singlet fission in DP-Mes was revealed, and we constructed a 'movie' of the full structural changes of the molecule over the course of a picosecond following photoexcitation. We found that two distinct kinds of vibrations—tuning and coupling modes—coordinate in a precise manner in order to enable singlet fission. In particular, tuning modes are responsible for bringing the initial and final states close to resonance, and coupling modes are only activated for a short time in order to enable population transfer. We thus showed that the singlet fission efficiency is extremely sensitive to the interaction magnitude of excitons to coupling modes. Indeed by artificially changing the coupling mode strength by $\pm 5\%$, we induced changes of up to 37% in the singlet exciton decay time, in the positive or negative direction. In practice, such changes could be introduced e.g. through the substitution of atoms participating in these vibrational motions, highlighting the practical potential of this approach. Of course it is not possible to try the millions of possible structural changes in the laboratory, however the very low computational cost of our tensor network methodology could be utilised for future high-throughput studies that would guide experiment towards a small and promising number of candidate molecules that optimise processes of interest.

Finally, in Chap. 9 we combined the physical insights from previous chapters, particularly Chaps. 7 and 8, in order to build a simple model for endothermic singlet fission in DT-Mes. This model accounts for the effects of excess energy excitation, the dielectric environment of the molecules, vibrational relaxation towards low-frequency modes and also the different role of tuning and coupling modes. We used our model in order to interpret complex experimental observations, and we revealed that depending on the polarity of the dielectric environment, singlet fission can occur coherently or incoherently. We thus presented the first well-controlled switch between these distinct regimes of quantum dynamics, which allowed us to tune the timescale of singlet exciton fission within one order of magnitude. This demonstration of efficient endothermic singlet fission that is accessible at will relaxes the requirement that $E(S_1) > 2 \cdot E(T_1)$, allowing for the possibility of a greater range of efficient fission materials. The dielectric environment offers but one way of tuning the mechanism of this process, and alternatives such as electric fields or strong coupling to light could be used to achieve similar control.

Overall this thesis has provided a comprehensive picture of exciton-photon and exciton-vibration interactions in organic semiconductors. This was enabled by a number of methodologies we developed for modelling these strong interactions, both in and out of equilibrium, going beyond perturbation theory and the Born–Oppenheimer approximation. These methods exceed what had thus far been the state of the art in this field, allowing us to draw several interesting conclusions regarding the physics of processes relevant to excitons and light-harvesting, and to formulate design principles for increasing their efficiencies. The methods and physical insights which appear in this thesis open a number of potential avenues for future research, some of which were briefly discussed in the various results chapters, and a few of which we briefly consider here.

In Chap. 6 we presented a Monte Carlo method for capturing exciton-vibration interactions in periodic molecular crystals to all orders, providing an accurate estimate of exciton energies at equilibrium through the inclusion of zero-point motion. One could extend this approach to isolated molecular systems, by coupling the Monte Carlo sampling technique to quantum chemistry calculations for excited states. This could potentially lead to a systematic improvement in the predictive power of excited state calculations in molecules. The author of this thesis is already pursuing this research avenue [1].

While the methods of Chap. 6 capture exciton-vibration interactions in molecular solids to all orders, this is only at equilibrium. In order to accurately study the out-of-equilibrium dynamics of excitons and phonons in these systems, one could apply the tensor network methods that were used in Chaps. 7 and 8 in the context of individual molecules, to periodic solids. In practice, this would involve an extension of the linear vibronic Hamiltonian of Eq. (7.4) to include an additional sum over q points in the Brillouin zone. Elucidating the fully non-perturbative exciton-vibration dynamics that result from photoexcitation of a periodic solid of molecules would be a critical step towards practical application, since optoelectronic devices are typically based on solid state films and not isolated molecules. Another critical step to take would be to extend the methods of performing quantum dynamics to finite temperatures. While there are ways in the literature for performing tensor network dynamics at temperatures beyond $T = 0\,\mathrm{K}$ [2], these do not allow one to extract observables of the 'bath', which in our case represents the vibrational environment and the real-time molecular/crystal motion. One potential way of approaching this problem would be to create displaced configurations of the studied system, which are distributed according to the harmonic density at the temperature of choice, Eq. (6.3), from which the quantum dynamics would be initialised and the observables of interest would be extracted. Then, similar to how we performed a Monte Carlo sampling of the integral of Eq. (6.2) for the exciton energy in Chap. 6, we could do the same for any observable arising from the quantum dynamics.

Apart from explicitly treating a torsional motion in Chap. 9, this thesis has entirely relied on the harmonic approximation for vibrational modes. However, low-frequency modes in molecular crystals are known to exhibit strong anharmonicity [3], as we have also highlighted in Fig. 6.11 for the case of a supercell of anthracene. This anharmonicity was the reason we excluded vibrations below $100\,\mathrm{cm}^{-1}$ from our description of ultrafast dynamics in Chaps. 7 and 8. While this is a reasonable approximation at short timescales due to the slow periods of these modes, anharmonic modes certainly need to be accounted for in the most general case. Indeed, a recent study highlighted the importance of anharmonic low-frequency motions in the real-time dynamics of singlet exciton fission in solid state organic semiconductors [4]. At the same, Fig. 6.11 and the associated discussion suggest that it is imperative to account for anharmonicity in order to capture long-wavelength phonon effects on the equilibrium properties of these materials, such as the exciton energies. One way that this problem could be approached would be by replacing our Monte Carlo sampling by a sampling of trajectories that are obtained from path integral molecular dynamics [5], as these simulations do not rely on the harmonic approximation and also include the effects of zero-point motion which we found to be critical.

The presented methods are general and could be used to study materials beyond organic semiconductors, where strong interactions between light, excitons and vibrations are present. One class of materials to which we believe our methods to be particularly applicable are two-dimensional transition metal dichalcogenides, which host tightly bound excitons [6, 7] that couple very strongly to phonons [8, 9], and strong coupling to photons has also been demonstrated [10]. Two-dimensional structures have emerged in recent years as promising candidates for applications due to the capabilities that their low dimensionality opens, with an example being stacking layers of different materials and to create heterostructures. These heterostructures have been suggested to have the potential to lead to exotic properties such as room-temperature superconductivity and exciton superfluidity [11]. We therefore believe that the task of uncovering the novel physics that the low-dimensional character of these materials opens is a highly interesting one, to which we hope that the work presented in this thesis will contribute.

References

1. Hele TJH, Monserrat B, Alvertis AM (2021) Systematic improvement of molecular excited state calculations by inclusion of nuclear quantum motion: a mode-resolved picture and the effect of molecular size. J Chem Phys 154:244109
2. Tamascelli D, Smirne A, Lim J, Huelga SF, Plenio MB (2019) Efficient simulation of finite-temperature open quantum systems. Phys Rev Lett 123:090402
3. Raimbault N, Athavale V, Rossi M (2019) An harmonic effects in the low-frequency vibrational modes of aspirin and paracetamol crystals. Phys Rev Mater 3:1–11
4. Seiler H, Krynski M, Zahn D, Hammer S, Windsor YW, Vasileiadis T, Pflaum J, Ernstorfer R, Rossi M, Schwoerer H (2020) Nuclear dynamics of singlet exciton fission: a direct observation in pentacene single crystals. https://www.science.org/doi/full/10.1126/sciadv.abg0869
5. Kapil V, Rossi M, Marsalek O, Petraglia R, Litman Y, SpuraT, Cheng B, Cuzzocrea A, Meißner RH, Wilkins DM, Helfrecht BA, Juda P, Bienvenue SP, Fang W, Kessler J, Poltavsky I, Vandenbrande S, Wieme J, Corminboeuf C, Kühne TD, Manolopoulos DE, Markland TE, Richardson JO, Tkatchenko A, Tribello GA, Van Speybroeck V, Ceriotti M (2019) i-PI 2.0: a universal force engine for advanced molecular simulations. Comput Phys Commun 236:214–223
6. Mak KF, Lee C, Hone J, Shan J, Heinz TF (2010) Atomically thin MoS_2: a new direct-gap semiconductor. Phys Rev Lett 105:136805
7. Wang G, Chernikov A, Glazov MM, Heinz TF (2018) Colloquium: excitons in atomically thin transition metal dichalcogenides. Rev Mod Phys 90:21001
8. Shree S, Semina M, Robert C, Han B, Amand T, Balocchi A, Manca M, Courtade E, Marie X, Taniguchi T, Watanabe K, Glazov MM, Urbaszek B (2018) Observation of exciton-phonon coupling in $MoSe_2$ monolayers. Phys Rev B 98:035302
9. Du L, Liao M, Tang J, Zhang Q, Yu H, Yang R, Watanabe K, Taniguchi T, Shi D, Zhang Q, Zhang G (2018) Strongly enhanced exciton-phonon coupling in two-dimensional WSe_2. Phys Rev B 97:235145

10. Dufferwiel S, Schwarz S, Withers F, Trichet AAP, Li F, Sich M, Del Pozo-Zamudio O, Clark C, Nalitov A, Solnyshkov DD, Malpuech G, Novoselov KS, Smith JM, Skolnick MS, Krizhanovskii DN, Tartakovskii AI (2015) Exciton-polaritons in van der Waals heterostructures embedded in tunable microcavities. Nat Commun 6:8579
11. Geim AK, Grigorieva IV (2013) Van der Waals heterostructures. Nature 499:419